To Betty,
A firefighter at heart!
Kathy Gillette

Firefighterette Gillette

A Firefighter Who Crashed & Burned

______________**A Success Story**__________________

Engineer Kathy Gillette

Alacheri Publishing, LLC

Indianapolis, Indiana

Alacheri Publishing, LLC
PO Box 26587
Indianapolis, Indiana 46226 USA
317 755 6670
www.alacheripublishing.com
www.firefighterettegillette.com

SAN 853-4616

Library of Congress Cataloging-in-Publication Data

Gillette, Kathy, 1954-
Firefighterette Gillette : a firefighter who crashed & burned--a success story / Kathy Gillette.
p. cm.
Includes bibliographical references and index.
ISBN 978-1-60348-022-2 (alk. paper)
1. Gillette, Kathy, 1954- 2. Women firefighters--United States--Biography. 3. Fire fighters--United States--Biography. 4. Fire extinction--Indiana--Indianapolis. I. Title. II. Title: Firefighter who crashed & burned--a success story.

TH9118.G55A3 2007
363.37092--dc22
[B]
2007022830

Printed in the United States of America
10 9 8 7 6 5 4 3 2 1

Dedication

To those of us who dedicate our lives to others, especially firefighters, police officers and EMS personnel, all of us on the street who come to your house in your time of need, think of us in our time of need.

Thanks Russell Grunden for your wonderful editing suggestions. This book greatly improved with your help.

Twenty Years!!! Belated Congratulations "Heinz 57", Indianapolis Fire Department Recruit Class of 1985.

Mandy, I love you. Best of luck on your retirement! I'll be joining you soon. My official retirement date is March 9, 2010. Mary, when are you going to join us?

Thanks "Women in Fire Service" for the great job you do. Keep on fighting.

Gary, everyone in our class wishes we had known you were so sad. We miss you.

This true story was written in my 19th and 20th year of firefighting. It is not necessarily in chronological order and some statements of fact may be incorrect due to poor recall. I never took notes in my years on the department, contrary to rumor.

Table of Contents

I.	Foreword	9
II.	Introduction	13
III.	Part One-The Academy	
Chapter 1	After high school	17
Chapter 2	Returning sweaty	23
Chapter 3	The next week	27
Chapter 4	We were soon introduced	33
Chapter 5	Some stability settled	39
Chapter 6	One morning we came	45
Chapter 7	The house to be burned	51
IV.	Part Two-The Real Fire Department	
Chapter 8	I was so excited	63
Chapter 9	The next day	73
Chapter 10	The mother house	79
Chapter 11	It's difficult for me	83
Chapter 12	My next big adventure	95
Chapter 13	Substituting at all	101
Chapter 14	When my certification	113
Chapter 15	Frequently in the evenings	119
Chapter 16	I got orders	133
Chapter 17	The next station	137
V.	Part Three-Going Regular-A Home	
Chapter 18	Going regular three	147
Chapter 19	In our district	157
Chapter 20	My screamer chief	163
Chapter 21	It soon felt	169
Chapter 22	Some of the new women	173
VI.	Part Four-A New Home	
Chapter 23	My new firehouse	183
Chapter 24	Around 0200 early morning	189
Chapter 25	Sometimes we shook	195

Chapter 26 The good old chief's aide 207
Chapter 27 Rumor reached us 219
Chapter 28 On Sundays I loved 227
VII. Part Five-Out of the Pan-Into the Fire
Chapter 29 My young neighborhood 241
Chapter 30 Released by my doctor 257
VIII. Part Six-Down the Homestretch
Chapter 31 Not long after 273
Chapter 32 Predictably the process 277
Chapter 33 I sobered up 289
IX. Afterword 297
X. Author's note in the interest of my ex 298
XI. Glossary excerpts from IFD training manual 299
XII. IFD Slang term excerpts from training manual 303
XIII. Author Biography 305
XIV. "Happy Sad Firefighter Stories" Book on CD and "Firefighterette Gillette" Exercise CD's ordering information 307
XV. About the cover picture 309

Foreword

It is truly astounding that two people, separated by over 500 miles and an international border, could have such similar experiences in careers that span twenty years. As I read this engaging and incredible story, I kept thinking, "Yep, that happened to me.", over and over again. From Kathy's military service, to her various jobs on the fire department, to her battles with health problems; I experienced all of these situations to some degree or another.

I spent time in the Navy and then joined the Toronto Fire Department in 1989 while I was attending university. When I started the Fire Academy training, we had instructors very similar to the ones described in *Firefighterette Gillette*. I was assigned a position at my first fire hall and met incredible people, some devils and some angels, but all of them were definitely "characters" in their own way. As I spent time in the hall (and time does indeed go by very fast), I, like all Toronto firefighters, got to try different jobs – engineer, hazmat, chief's driver – these roles give us a better overall education and understanding of what the fire service is about.

Being a firefighter, I always like to hear fire stories – they are both instructive and easy to imagine. To properly understand the chaos that rages inside a house fire, you have to experience it. I refer to it as a "rumble in the dark" because it sounds like a gang fight that is occurring in the structure. Due to the pitch blackness – people are banging walls, bumping into each other, screaming through their facepieces, and throwing around furniture, knocking it out of the way. It is a wonder that the job gets done. When I think about some of the fires I battled, I am impressed with how everybody pulls together to extinguish the blaze, considering every fire is very different.

When I first worked on the Hazmat truck, we attended all second alarm fires in the entire city – which is basically every working fire. The one I'll never forget was a huge blaze that occurred on a Sunday afternoon at an east end chemical plant. This business stored many petroleum resins and tars. Its facilities were filled with driveway sealers and roofing materials. As the warehouse fire grew larger and spread to nearby buildings, people from miles away could see the plume of black smoke – it was truly a spectacle. Up close, the smoke became coloured with purple and green tendrils as different chemicals ignited to add to the blaze and, to complete the picture, random propane cylinders exploded as they became too hot, whistling through the plumes.

The devastation continued through the night with many crews stuck in their positions for hours at a time becoming exhausted while commanders struggled to keep up with the demand for fresh bodies. The fire was finally extinguished just before midnight by a concentrated effort of foam attack lines. Foam was not considered earlier because of the requirement for copious amounts of the concentrate which was unavailable. We later discovered the chemical plant next door had quite a lot of it and put it to use or we might have been daunted by the ferociousness of the fire well into the morning hours.

A few years later, at an incident in the west end of Toronto, crews were searching for a fire below a variety store. Crawling around, looking for the basement door which seemed impossible to find, firefighters contended with extremely hot, smoky conditions. As the entry was located, I arrived with my truck, and as luck would have it, I was in the captain's seat that day. As soon as we announced our arrival, we were tasked with bringing a 65 mm hose through the basement door to proceed with an attack. I wondered if this fire was going to be extinguishable by us, now that it had plenty of time to become well-seated. Proceeding to cover our exposed skin with every available flap we had on our gear, a fellow coming out had warned us of the intense heat, we were given only vague directions as to where this door was and directed inside to confront the fire. Someone announced there was a hole in the floor which complicated the bleak situation.

As we entered, crawling on our bellies, we were hit with the most intense heat I have ever experienced. Even though I had on my flashhood and helmet flaps down, I could feel my ears burning after we were just two feet inside the doorway. We inched along in the general direction of the door but I could tell this place was getting way too hot to stay and decided for the safety of the crew to retreat. Before I could even tell the crew my decision, someone else reported the floor becoming spongy, and the district chief radioed for everyone to get out. We fought the fire from inside the doorway with flames shooting out of the storefront right above us.

After the conflagration was brought under control, personnel reported injuries from the flashover that occurred barely seconds after we left the building. I received 2nd degree burns to my ears which took a couple of weeks to heal. It was the only time I have ever been burnt at a call. The chemical plant fire created long term casualties - a number of people have recurring health problems including a handful of cases of testicular cancer (many of these men were up to their waists in runoff water for extended periods of time). As you will read in Kathy's account, this occupation is not for the faint of heart.

Kathy relates some amazing stories – calls that I would love to have been at and a few that I am glad I wasn't. The part that really engrossed me, however, was all the fire hall politics that she had to contend with throughout her career. The only real difference between the two of us (and it is an important one) is our gender. I never had the same proving ground that Kathy had to endure. I was impressed by her tenacity at grasping new opportunities and her ability to overcome the large obstacles put in her way.

I was greatly interested in the differences between our respective jobs; we have some divergent policies that impact how our departments function. Kathy was initially assigned to a "sub" position which basically means that she had to fill-in wherever a vacancy was found. In Toronto, we are placed in a permanent spot on a specific truck from day one and we share, as a group, the requirement to fill-in (more or less). Kathy had to apply to become an engineer (the driver) and she was given extra money once she succeeded. Toronto firefighters share the duties of driving and, unfortunately, we don't get paid for it. Kathy was able to take paramedic training – our paramedics are separate from the fire service. Kathy's promotion seemed to be held up for personal reasons. Our system is based on seniority. Once qualified, firefighters remain in the process until they reach the top and get promoted.

The greatest difference I noted was not directly related to firefighting but, concerned how medical calls were conducted. In Canada, health care is provided free of charge by the government and paid for through taxes. We don't have the task of convincing someone to go to a hospital because they cannot afford to go and feeling we have given inadequate care because we know the citizen signed the release form because he did not want to face the hospital bills. Everybody receives the exact same care and when they are feeling better, they leave the hospital without worrying about paying a bill. I would not want to be a medic and have to deal with whether or not a patient will die because he or she cannot afford the treatment.

In closing, *Firefighterette Gillette* is truly addictive. I brought it everywhere as I was reading it. Any time I had ten minutes to spare, I read another chapter, looking forward to the next opportunity to continue the story. I anxiously await more from Kathy (I am sure there are still some tales to tell) and I am honoured that she allowed me to present her amazing book for all of you.

Graham Voss
Toronto Fire Services
Canada

Introduction

Too many memories of the last 20 years are fading. Some are nightmarishly clear, the memories of the dead ones, especially the children. When I allow myself to think about them, I feel a lump developing in my throat and the hair rising on my arms, so I don't think about those times, much less talk about them. Some nights, those memories wake me, especially firehouse nights, and then I can't get back to sleep. If I am not awakened by their faces, then it is the alert that is waking me, mostly for medical runs, but sometimes fires. The night runs are hardest on firefighter hearts. Being wakened to burst into action all hours of the night for 20 or 30 years sends a lot of us to an early grave.

Night runs are also hell on kidneys. We jump out of bed into boots and nightpants, get on the fire apparatus, and on the way to the run, realize how bad we have to pee. This is a bigger problem for the women than the men. The older guys can hardly buckle their night pants giving them a head start on peeing. They rely on suspenders to hold up their unbuckled pants.

Wow! Enough of the negative stuff, because you are going to think I don't like my job. I love my job! I feel lucky and privileged to have served as a professional firefighter in my lifetime. It also feels good to have my 20 on and be eligible for retirement, although I probably won't retire until my husband has his 20 on, but maybe I will. Like Scarlett, I'll decide that tomorrow.

My handsome son, an adult now, was a baby when I was hired as a firefighter. Recently he rode with my crew and me. After the frantic excitement of barreling down the street with lights flashing and sirens blaring, he witnessed our cool, calm, organized reaction to a flopping, seizure patient on a store floor, where people were rushing out of the building spluttering, "She's crazy!"

The fire, which happened late that night, had flames leaping from every window. He observed as we quickly doused the blaze. Getting involved in the action, he helped pick up hose and met my chief.

Impressed, he decided he would like to follow in my footsteps.

"Oh my God, Mom, I never realized what you do!"

Anyway, I might be getting a little off track here. I should start at the beginning. I should start with my decision to become a firefighter.

Part One

The Academy

Chapter 1

After high school, I joined the army looking for excitement and fulfillment. Assigned a desk job in the Women's Army Corps, I found I was stuck in a job that was neither. Wanting to keep up with other soldiers, I learned to swear like mad, one skill that served me well in the fire department. I stayed six months past my required enlistment and said good-bye to army life. After graduating from college, I worked in the corporate world, where I still wasn't satisfied. Steve, my older brother, who joined the fire department a year earlier, told me his adventures as a firefighter. My dog eat dog sales job seemed lonely and boring compared to his exciting, roller coaster, job of firefighting. Missing the camaraderie of the army, I began to hope I might find that component in the fire department, minus all the bullshit.

My brother warned me, I would run into some male chauvinists who believe the last place women belong is in the ranks of the fire department. Being the big, women's libber that I am and hard headed to boot I decided to take my chances. I applied and nine months later, (the application process is that long, and sometimes longer) gave birth to a job as a probationary recruit on the Indianapolis Fire Department. At our swearing in ceremony, (my memory is hazy as to what exactly we swore, to honor and protect or something like that) we stood and raised our right hand, or put it over our left breast and repeated our oath after the public safety officer. The mayor gave a speech about how proud he was of our class. With cameras flashing, he announced the four women in this class of 32 would double the number of women firefighters in the department.

At the time, although I was snickering under my breath, I didn't realize how truly progressive this mayor was, because after 25 years of women being on the job, we only have 35 women fighting fire and 700 men. The 2006 class had one woman in 46 recruits. I heard through the rumor mill she was hired because she is the wife of a chief on another department who is friends with the chief on our department, and he asked for a favor. That is what I heard from a fairly reliable source, another firefighter. Maybe you've heard the old saying...telegraph, telephone, tell a firefighter. Nothing is private on the fire department. The first page the guys read in the newspaper is the divorce and death columns. Every firehouse gets at least one newspaper, some get two.

The next day, all of us recruits reported for class, 8:00 a.m. and you'd better be early to the training academy on Post Road, which is an old converted school also used by the police department and other government agencies. It was a relatively new facility to the fire department. The training tower wasn't completed yet, so our training was divided between there, and an old tower on Michigan St., or maybe 10th St. Due to budget constraints, more recent classes have been trained at other facilities where it is cheaper but not as convenient.

The first thing on their agenda was to round up the women recruits to tell us to cut our hair off, tonight, or don't come back. We were given a copy of the hair regulations and warned to ensure our hair was within the guidelines. The instructor said the female's picture in the general order was not a flattering one, but she is nice looking, even with short hair. I was surprised by this order, because I was in the US Army for 3½ years, part of it during the Vietnam War, and only had to put my hair up. My hair had been long most of my life. I didn't want to cut it, but there was no turning back now, because I had a husband to put through college, a baby to feed, and a mortgage to pay. My husband had two years of college left. There was nothing to do but follow orders which I knew how to do.

We met our four main instructors who would be with us throughout training. We had an engine instructor, a ladder instructor, a physical fitness trainer, and the director of the academy. We also had visitors from the local union, who signed us up as members. We introduced ourselves, and added a tidbit of information. I announced that I was one of the oldest in the class, and had a 1½-year-old baby.

After introductions, we were given a set of rules, or general orders, and told, "Read them tonight and be prepared for a test in the morning."

Finally, we got a break for lunch. I was starving, but that is my general state of being. I brought my lunch and ate in the small break room, where some vending machines were stocked with sodas and cupcakes. Bottled water hadn't arrived yet, nor had the fire department begun a nutritional program. Work out and pump up with junk food. There were quite a few budget conscious recruits with me until they got their first paycheck.

After lunch, we were encouraged to sign up for "Death and Retirement". That is a club on the fire department where each time a member retires, or dies, they or their family receive a lump-sum payment from all of the contributors, the more people in, the more money a firefighter is gifted. Just a few years before, the operator of the program, a firefighter, was caught stealing big bucks from the fund. Most of the money was not recovered from the perpetrator, nor would it ever be. Since the scandal, the program suffered from some dropouts of the younger firefighters. My brother cautioned me not to join due to the fund's instability.

After the induction procedure was finished, I was the only one who had not signed on the dotted line.

They called me up front and asked questions like, "What is the problem?" and, "Don't you want to be a team player? This program is so everybody gets a gift from the fire department upon retiring, not just the chiefs, and the poor widows usually need the money for funeral costs."

My questions about solvency and security weren't addressed, so I declined again. I was told to sit down. If they had pressured me a little more, I might have caved-in, but the tightwad in me could handle that small amount of embarrassment.

We were lectured the rest of the day, kept five minutes late, then released to read our orders. I drove the 40 minutes home in rush hour traffic, cooked dinner, played with the baby, went to my mom's house around the corner and had her cut off my hair. I went back home to bed without cracking the general orders book. Since my husband's college classes started later, and let out earlier, he was put in charge of getting the baby to and from my mom's house.

I packed some shorts, t-shirts and towels for our second day of class, which started with physical training (P.T.). Because we wanted to keep our jobs, and lateness was absolutely not tolerated, everybody was early, standing around in the hallway or break room, waiting to file into the classroom. We set our workout clothes by our seats that were assigned to us the day before. Promptly at 8:00 a.m., with everyone seated, attendance was called in alphabetical order.

We sat four to a table. I was a "J" and had half a class in front of me all responding, "Here, sir."

For kicks, and because I was still upset about being forced to cut my hair, I responded, "Here."

There was a pause, then the yelling started, "Did you not read your general orders last night as instructed, Private Jameson?"

"Yes, I did," I lied.

"You didn't read them very well, because the orders state you are to address all of your superiors as sir, and we are your superiors, Private."

"Yes, sir."

Well, that was fun. I thought I better get a little more serious, since they didn't seem to have any sense of humor at all.

After making an ass of myself, which I still do sometimes, and the rest of the class answered, "Yes, sir, here, sir." we filed out for P.T.

Since we were in an old public school, there were lockers for women. The four of us were getting dressed and chatting quietly since we didn't know each other very well. When the woman next to me took off her pants and turned around to speak, I saw her underwear had a huge hole. Half of her pubic hair was hanging out. We all stopped and were speechless, staring at the hole in her panties. She started laughing and explained that she was broke. She hadn't been able to buy new underwear. She bought new socks, but forgot about her worn underwear and the need to disrobe in front of people. She planned to get new undies with her first paycheck. We would not have to look at her "holey panties" anymore.

I have always loved sports and played basketball until I dropped, but I had a baby, whom I was still nursing and hadn't quite lost all of my baby fat. I ran to get in shape for the agility test, but that had been six months prior. I hadn't done much since then because I was busy with so many things. A baby is time consuming.

Our instructor worked us hard. First, he got baselines on our vitals: heart rate, respiratory rate and blood pressure. He pinched us for body fat. He also tested us for weight lifting capacity. My instructor was not impressed with my scores. He said he had never seen anybody with so much fat (the other recruits are staring at me now), and lifting so little weight. I felt wholly inadequate.

Then we all ran. I wasn't in the lead, but I didn't bring up the rear either. I was buried in the middle, which made me feel a little better. To end our training for the day, we were timed doing sit-ups and push-ups. I did the fifth most sit-ups and finished in the middle for push-ups.

As we headed for the shower, the holey underwear woman, Mandy, tapped me on the shoulder and said, "I heard what our instructor said to you about being fat and lifting. For a woman, 20 percent body fat is normal, and hardly any of those guys got anywhere close to lifting their body weight. Proportionally, you are about the same as everybody else."

She used to work in a gym and knew all that “meat stuff”. She had even done some competing and that was why she out-lifted most of the guys. Lucky for me, she rescued my deflated ego. I now believed it was just a matter of training and conditioning. I knew I could get into shape. I would just train until I dropped.

Chapter 2

Returning sweaty from physical training, we took our seats in the classroom. It was March and the cool room felt good. We studied so many things out of fire books. We learned about construction and how it collapses in fires. We contemplated the history of fire, and why it is considerably more toxic today, due to plastics. We memorized some of the most dangerous fire gases, they being the reason we should always wear our masks into fires, even though we will be taunted in the companies because we aren't true smoke eaters.

The director of the academy, an acting chief, (he was a captain filling in for a chief's position) told us how he suffers frequently from upper respiratory infections due to inhaling toxic gases over the years and he does not want that to happen to us. We could be messengers to the old guys about fire safety, and prolong our lives by wearing masks. During his speech, his beeper went off. He said there was a working fire close by and didn't want to miss it. He got our engine instructor to take his place. Off he went to run into the fire.

The engine instructor appeared with food between his teeth, having had his lunch interrupted. He said he was not prepared to give us this lecture, but since he didn't have any of these, flicking his shirt collar where officers wear their bars, he would have to follow orders and make something up. "I don't have these!" and the flicking motion, became a familiar, every day, overused saying. Later the captain returned with soot and ashes all over his ears. He had hurriedly washed his face, but must not have looked in the mirror.

My mom's words came to me, "Be sure and wash behind your ears." He missed that lecture from his mom.

He told us about an experiment he tried. He read a fire instructors magazine that discussed how most fires are "drowned out" with a lot of water damage to the structure. The author of this article believed fires could be "steamed out" resulting in a lot less damage to the structure. Our fire department was using fog nozzles (a stream of water that has lots of droplets, like a shower head, versus a straight stream, which is like what comes out of the bathtub spigot), ideal for "steaming out" fires. "As I was in the house by myself, with just the crackle and smell of the fire, I came to the bedroom door and carefully cracked it. I opened the fog nozzle and gave it 100 gallons of water directed at the ceiling. Now, you all know, that is one minute of water, right, because fog nozzles put out 100 gallons of water per minute. I shut the door for a few minutes letting the water dissipate into steam. When I opened the door, the fire was out, and there was minimal water damage."

My interest peaked and I thought, "There is a science to firefighting and even beyond the bravery, it is actually extremely amazing!"

We studied the fire triangle: fuel, heat and oxygen. Some feel there is a fourth element, chain reaction. Without one of the elements, there is no fire. Sometimes one of the elements has been used up and the rest are just waiting around. When the missing element is reintroduced, big problems may occur, like a backdraft.

Backdrafts happen in a confined space with superheated materials, and no oxygen. When someone or something introduces oxygen, BLAM, the place explodes into flames.

Flashovers occur when the room is heated to the point that the entire contents burst into flames, leaving us just seconds to bail, or burn. We were warned, due to the hot temperatures of plastic materials and our rapid response time, many firefighters are caught in these. Also, we go deeper into fires now, due to our equipment shielding us from the heat. We seldom realize how hot it is. When we do, it is sometimes too late.

After class, Mandy and I were talking. I admitted that it all sounded very daunting. Mandy said it was just scare tactics, and not to worry so much.

"Just listen to what they say. You never go in alone, but the captain says he was by himself, in the peace and quiet of the fire, just listening to it crackle. That is total bullshit. When are you going to be alone when two engines, a truck, squad and chief are dispatched together on small fires and even more are sent on bigger fires?"

The next day we were "invited" to an orientation meeting for our families. Held at the end of the week, we could invite whomever we wanted, but we had to be there along with at least one member of our family. I brought my husband, leaving the baby with Mom. I don't know what I was expecting but certainly not what was said.

Our families were informed, "Firefighting is one of the most dangerous jobs in the world. Your question should not be, 'Will my firefighter get hurt?' You need to know that your firefighters will most likely be injured several times in their career. They may even die or become permanently disabled."

Not many questions were asked after that. Slides were presented showing various fires and the hard work it takes to extinguish them. My husband looked a little pale. Stunned, everyone quietly shuffled out.

On the way home, my husband asked me, "Why do you want to do this? It is so dangerous! You could earn more money in the business world. Just how much of a pay cut did you take? Wasn't it about $9,000?"

I was flabbergasted. These words were being uttered by my husband, the man with whom I rappelled off mountains, and raced in downhill skiing. We met in the army where we played hard. I jumped out of barracks windows to sneak out with him. Isn't that sweet, he was worried about me. In retrospect, he was probably more concerned about the money and losing his meal ticket. I explained to him that the pay would catch up after a few years, and all this training would help keep me safe. We picked up the baby and went home for the weekend off.

Chapter 3

The next week started with more classes and lots of tests. I was reading the material now and doing well on the exams, considering most of this stuff was totally new to me. A couple of the guys had been firefighters on other departments, mostly volunteers. They could hardly stay awake during lectures, ordered to stand up in the back of the room if they nodded off. Others were struggling with the material. The instructors bullied the strugglers by firing questions at them and informing the class that we were being held up by a few deadbeats. The instructors said they wouldn't be worth hauling to a fire anyway, so they might as well wash out here and now.

We lost one or two recruits early in the game but most managed to hang on. It wasn't until we got to hazardous materials and studied chemistry that we lost three more recruits with quite a few in danger of flunking out. The instructor, an acting chief, did not help with his dry presentation of the material. One of the guys in our class majored in chemistry in college and organized a study group for everyone. I didn't go because I felt comfortable with the material but most everyone else went. It was amazing how a few hours with our fellow cadet clarified molecular structure for everyone when a few days with the chief had confused us.

The chief pounded in our heads, one of the most important concepts of hazardous material firefighting, "If you don't know, don't go; it might blow."

That saying and being encapsulated in rubber suits while trying to pick up dimes off the floor are most memorable. Bending over in those suits with the humpbacks where our breathing equipment is housed was quite a challenge. Picking up a dime with our fingers so covered in sweat under those rubber gloves that they looked like we had soaked in a bathtub for hours was practically impossible. Anyway, the "if you don't know..." saying comes from the necessity for proper PPE or proper personal protective equipment, and correct containment and extinguishment procedures. Some chemicals can have sinister reactions when mixed together. Water, put on certain metals, like magnesium, can cause horrendous explosions, so it is extremely important to know just what chemicals are involved before anyone rushes in and does something that makes the situation worse. It's a mortal sin to become a part of the problem instead of a solution to the problem.

My final memory of hazardous materials is a BLEVE (boiling liquid expanding vapor explosion). Thankfully, I have never been involved in a BLEVE (pronounced blevee), a terrible phenomenon. We saw many pictures of overturned tankers in a slide presentation given by our instructor. The tankers, full of volatile liquids, had BLEVE'd and the eruptions could be seen from miles away. The destruction around them was mind boggling.

Some time during this week we were taken to the quartermaster and issued uniforms. We were also fitted for our fire gear. The uniforms had to be worn to class the next day, but our fire gear would not be ready for months. We were scheduled for a lot of classroom training, so there was no rush. Our uniforms weren't bad. They were somewhat masculine but comfortable. Granny would have loved the shoes. I've never worried much about how I look so it didn't bother me. None of the women complained. We were pretty sure we'd be laughed at if we brought it up and told to leave if we did not like it.

Reporting the next day in uniform, we learned rescue techniques. First, we studied them in books. Then we practiced them. We utilized a lot of different carries. Most of them were for two people and they were drags. We practiced dragging each other around by our arms, legs, collars, anything grabbable. My favorite carry is the firefighter carry. You grab an arm and a leg, bend at the knees, and throw the person over your shoulders as you are rising up. If done in a swift motion, most of the person's weight is borne by your legs. Sometimes I do this at parties to entertain everyone. I always do it at schools to amaze the kids for career day. It gets a healthy laugh and a long line of kids wanting to be picked up and swung around. The biggest cheer goes up when I lift and twirl the embarrassed teacher.

After we learned all the carries, we were taken to an equipment room and handed fire gear donated to the academy by retiring firefighters. Needless to say, no females had retired yet so all of the gear was two, three, or four sizes too big for all of us women. The instructor insisted we wear it anyway and "do the best you can with it". We went back to the classroom for more book illustrations and studies of search and rescue techniques.

We then donned our salty, old, ill-fitting, fire gear, putting into practice right and left hand searches. The helmet I was given was so big on my head that even with the chin strap tight enough to practically choke me, the thing kept falling over my eyes every time I bent my head down. I kept my head up during the lecture but I knew I was going to have real problems in a minute because we were about to get down on our hands and knees to crawl around. The worst problem turned out to be my boots.

In smoke filled rooms, we can't see our hands in front of our faces so it is important to keep our bearings or we will get lost and trapped in the fire. One method of doing this, if we don't have a hose line to follow in and out, is to put a hand on the wall as we crawl. If we keep that same hand on the wall throughout our search, we will eventually come back to our exit.

We got down on all fours in one of the small rooms and began crawling in a right hand search pattern. Since we didn't have our air tanks and masks yet, most people could see, but not me. As soon as I got down, my helmet plopped over my eyes obscuring all but the floor below me. As we crawled, my boots started inching off my feet. I had to reach back and pull them up every few minutes. This slowed me down and a gap developed. Pretty soon, the gap became an attention getter and the instructor yelled at me, “Can’t you keep up, Private?”

“My boots are coming off, sir.”

“I don’t care what your problem is, you had better keep up.”

I crawled in earnest bumping into the butts in front of me whenever they slowed a little since I couldn’t see and then got behind when I would have to pull up my boots. The line wasn’t running very smoothly around me and I was taking a lot of abuse from the instructors. In an effort to go more smoothly, I quit pulling up my boots but then crawled right out of them. Leaving them, I kept going. The instructors stopped us and let us get up. I listened to the rest of the lecture in my stocking feet. Even though the instructors tried to act hard core about it, I was pretty sure they keeled over laughing in their office. I was very glad when that day was over.

The next day started with more studying. During our first break, I telephoned my parents’ house and was surprised that my dad answered. My sister had gone into labor with her first baby the night before. When I left the hospital late that night, she still had not delivered. I asked Dad why he wasn’t at work. He said things had gone terribly wrong with the baby’s delivery and the little girl had died. He was leaving for the hospital.

Hanging up, I sat down in a chair with my head between my knees because I thought I might pass out. Mandy asked me what was wrong, then gave me a cigarette. Not having smoked in a while, I became dizzier. (I quit smoking when I became pregnant.) One of my classmates got an instructor who wanted to know my problem. When I tried to speak, I started to cry. They took me into the office where I blubbered out the death of my niece. I was released from class for the next couple of days.

I went straight to the hospital to be with my sister who was drugged and could barely speak.

She said, “I did the best I could, Kathy.”

I brushed her hair away from her face and patted her while she cried. She said the cord was wrapped around the baby's neck which strangled her as she moved down the birth canal. When the baby's heart stopped, the doctor grabbed her with forceps and pulled her out, cutting the cord. The baby was revived after 20 minutes but died a few days later. Visiting her in intensive care, I marveled that she was so pretty, so perfect with the exception of her eyes which were milky white from oxygen deprivation. Everyone in my family was devastated. My brother-in-law and sister were inconsolable. After the funeral, I had to go back to school.

Everybody welcomed me back. They said I picked some good days to be gone and showed me all of the blisters on their knees. They had spent the past couple of days crawling, crawling, and crawling. They crawled through the gym, up the stairs, in the bleachers, over the chairs, under the desks, around the hall, past the bathrooms and back to the classroom. Then they started all over again in the opposite direction. They put the gear up at the end of the day and were moving on to the next subject. That was all right with me because I was in no mood to crawl out of my boots and bump into butts.

Chapter 4

We were soon introduced to the SCBA or self contained breathing apparatus. It would supply us with lifesaving air in fires. We wore the face piece portion while we ran in P.T. every morning. Still upset about my niece and feeling very down as we headed out to the track, I started crying. I ran very hard attempting to work out the tension. Between the crying and the sweating, I fogged up my mask and could barely see. One of the guys sidled up next to me and gently took my arm to help lead me around the track. Feeling a little claustrophobic, I was grateful for the help.

Why I felt that way now, since the hazardous material suit was much more confining, I'm not sure. The only thing I can figure is that the fog obscured my vision and I was frightened because I couldn't see. This was not a good trend to start. I tried to work through it by talking and listening to my friend guiding me along the track. The instructors were issuing these masks to us now so we could get used to wearing them. I thought the feeling of claustrophobia must be a fairly common malady with beginners like me because why else would we need to get used to wearing the mask. I comforted myself with the knowledge that others had overcome it and I could too. Thank heavens I was getting an early start.

My time in the two mile run was improving dramatically. I moved up in the pack. For exercise at home, I put my son in a bike seat and rode around the neighborhood to warm up. I would then hop off the bike and push it around the block. My son really enjoyed this and I was becoming very fit again. It definitely takes work to get back into shape after having a baby, especially when you put on 60 pounds during pregnancy.

Even though I was improving physically, my mind, contending with the grief, was functioning poorly. I could not seem to concentrate and received my lowest test scores. Although my academic ranking lowered, I was still in the top ten. That felt good considering I was competing with several ex-firefighter volunteers. I was in no danger of flunking out like some. We began preparations for emergency medical technician (EMT) training which our instructors warned us would be intense and difficult, and probably knock out the stragglers.

That was okay with the instructors though because, "If they couldn't make it through this, they wouldn't be worth having on the EMS (Emergency Medical Service) runs anyway."

For EMT class, we were issued a thick book. A couple of the guys were paramedics, far more advanced than EMT's, but they still had to attend class and take all of the tests. This reminded me of the stupid things the army required of soldiers. The nit picking things the instructors did and the "we want to wash you out" attitudes were getting on my last nerve.

I griped to Mandy, "And I cut my hair for this."

I accusingly questioned my brother why he would recommend a job like this. He told me the academy sucks but being in the companies was the greatest and I should just hang in there until I was through with the idiots in the academy. He assured me I would love the job once training was completed.

I wasn't the only recruit grumbling.

One guy, who had been a teacher before he came on, said the atmosphere was not conducive to learning. "If the true spirit was to teach, then they should change their instructing methods."

Therein was the key. The objectives were to teach and to stress, breaking those who could not hold up under pressure. I doubled my resolve. I wasn't going to crack after my brother's reassurance.

So began EMT training and my renewed sense of urgency to get through the material. Read, read, and read more. We were given lots of reading assignments each night and then reviewed the material in class. Along with the lectures, we practiced the basic medical skills we would perform in the companies. The instructors said they wanted to build proficiency in these skills because we would be the backbone of the department in the coming years.

The older guys were not happy about being forced to take EMS runs. The administration and the union said it was the future of the fire department and would save our jobs when the city was looking for places to cut. The public safety office had recently closed several fire houses and consolidated others in order to minimize costs. Several hundred jobs were lost to attrition. It had been quite a few years since anyone had been hired. The guys in the classes before us were roving for five and six years as substitutes at stations when someone was sick or on vacation. Some who were scheduled to become regular (receive an assignment at a firehouse where they would be a permanent member until they wish to move elsewhere) had been roving for seven years. The rovers were pulling for us to graduate, because for every recruit that drops out, one less person gets to go regular.

A number of the older guys on the department refused to become EMTs. A six hundred dollar annual bonus for being one was not incentive enough to make them want to go on runs for sick and injured people. They believed firefighting was their true profession and anything else was too mundane for them. They feel EMS compared to firefighting is like high school football compared to professional football. They both have similar rules but it's a different ballgame.

These guys came on when there was nothing but firefighting and didn't ask for or want the medical component. It was shoved down their throats. They were told to leave if they didn't like it. They had one option, move to an apparatus that didn't require EMS, a truck. The main differences between an engine and a truck (sometimes called aerial) are: an engine carries hose and water, a truck carries ladders and heavy equipment. Truck people knock down doors and chop holes in roofs; engine people put the wet stuff on the red stuff and help sick people. More truck people than engine people get killed because they run into burning buildings without water to protect themselves. You can't beat back much fire with an axe or chain saw. Quite a few stubborn old men were on the trucks. (Eventually all truck positions had to be filled with EMTs and the last of the hold outs retired.)

Back to the practicals in EMT class, we learned vitals are important. You can tell a lot from blood pressure, breathing rate and heart rate. If they are not within normal limits (wnl), you better start asking questions. We practiced taking each other's blood pressure and pulse. They seem simple but they do require some experience.

Our class was divided into tables of four. As people dropped out, empty chairs were left at the tables. Since my table was still full, we took blood pressures and pulses right there. My tablemate was grumbling that IFD had no paramedic program. He said he was investigating other departments that offered it and was considering leaving for a township department where it was available.

He took my arm and said, "I hate taking BPs on people with flabby arms."

I don't know what was up his ass but just because the instructor was calling me fat, I wouldn't let him. I jerked my arm away and told him I was not flabby and he could just shut the fuck up. He resigned a week later to join the other department. I haven't seen him since.

In other practicals, we took turns being victims of: broken bones, contusions, abrasions, burns, lacerations, heart attacks, brain attacks (strokes), single casualties, multiple casualties, tied up to backboards, splints on every limb, slings on arms, bandages over our eyes, rapid transports, on scene stabilizations, etc. We familiarized ourselves with the bag valve mask, a kit that has a face piece connected to a squeeze bag that can deliver 100 percent oxygen when augmented with 15 liters of oxygen per minute. You name a basic life support skill and we practiced it. We spent weeks fixing each other up. We were in earnest because in addition to oodles of written tests, we would have to show our new skills to perfect strangers who would grade us. If we didn't pass, we would have remedial training and retest. If we still didn't pass, we were out and a few more were out.

Chapter 5

Some stability settled over the class after those of us who passed EMT school prepared for fireground evolutions. Our brand new gear that actually fit arrived, all except the helmets which were on back order. I managed to find one that stayed on my head and kept track of it so I could beg for it if someone else got it first.

The first couple of weeks of fireground operations were devoted to technique so everything was done at a reasonable pace. A skill was demonstrated and then practiced. We were shown several ways to don our SCBAs, allowed to practice the different techniques and then choose the method we liked best.

An interesting exercise was the "Church Raise". A ladder, maybe 24-feet tall, was tied with four ropes at the top and then raised. The ladder was positioned in the middle of an empty parking lot. Four people were placed at the ends of the ropes supporting it. We then climbed up and over the ladder and down the other side. We were told this would build confidence in our classmates. When I told my brother about the confidence builder, he warned me to beware. His confidence-in-your-classmate-building-session ended abruptly.

His classmates held a safety net at the bottom of the three story tower. He jumped into the net and they dropped him. Luckily, he landed on his helmet which cracked in half leaving his head intact but with a big headache. They suspended the operation forever in class and on the firegrounds. My brother was the last person to ever jump into the net around these parts.

After we were shown how to: set up ladders, lay every size hose, chop holes, set up fans to blow out smoke, instructed how to properly pull down ceilings with pike poles, etc.; they brought in the engines with the water.

We learned how to turn on the hydrants and "throw" water. Not having a clue about hydrants, (The instructors never forget the stupid, irritating questions that I asked the local water company, "What do you mean you have 37½ hydrants per square mile? How can you have ½ a hydrant?" Whoever said there is no such thing as a dumb question, never heard that one.) I was ordered to be the first volunteer to open one. I was given a long handled wrench that could split a skull with one swing. Attaching it to the nut on top, I started turning the screw making a complete circle around the hydrant.

The instructor screamed at me to stop, "You look like a mule bound to a grinding wheel."

He showed the proper way, which is to plant your feet and pull the wrench around, making it a speedy operation. Fast is good when getting water to the apparatus because one hundred gallons of water being poured out the end of a nozzle on a fire inside a burning building gives the engineer outside just five minutes to get more water from a hydrant before the tank of five hundred gallons is emptied, leaving the firefighters inside without water. Whew! That means you had better turn on the hydrant as fast as possible instead of slow, like a mule. Thanks to me, everybody knew how to open a hydrant without looking like an idiot.

All facets of the operation were put together. A gang of us would get on an engine at one end of the parking lot. We would ride the engine to the hydrant, all of us hanging on to whatever we could grab. One designated person would hop off to pull the hose and wrap it around the hydrant and signal the engineer to take off. The engineer was an old guy who was either detailed or volunteered to help. He would drive another 50 feet to the tower where we jumped off with our various truck and engine duties to complete. We were taught never to run because with the extra 60 pounds we were wearing, ankles are easy to twist. We were definitely hustling as if we were in a race. If we didn't hurry, the instructors were calling us a load and screaming in our ear.

The injuries started piling up. Mandy was standing on a hose when it was jerked out from under her. She fell on her wrist, was whisked off to the doctor and came back in a cast. Two others came back on crutches, Mary with a twisted knee and one of the men with a spiral leg fracture. Another had a pulled groin muscle. Even though the injuries were mounting, the instructors pushed us with feverish intensity.

There were near casualties. Several of them involved me. Our instructors liked to engage us in competitions. Sometimes the instructors didn't plan things very well or maybe it was just what they had in mind. Two teams hooked up to the engine's outlets, one from the right and the other from the left. We dragged the hose around opposite sides of the tower. When we came in the door, we got down and crawled like we were in a fire. When we rounded the corners, we were ordered to open our lines. I was on the front of the nozzle and when the other team opened their nozzle, they lost control and shot water in our faces. We were only about five feet apart. The power of the water tore off my cinched helmet and blinded me. The instructors shouted for the water to be turned off at the engine. They helped me up and lead me outside. After a few minutes, I could barely open my eyes but everything was blurry. I got to sit out the next half hour until lunch break. My instructors gave me an A for the day because I hadn't dropped my nozzle, even under duress.

We spent a day learning to tie knots, all kinds of knots: square knots, bowline, running bowline, clove hitch, half hitch, slip knot, maybe others that I can't remember. We used these knots to secure the ropes on the ladders and to tie ropes onto the equipment which we hauled up to the third story window. We raised and lowered equipment with the knotted ropes instead of running up and down the stairs.

It was my turn to lower equipment. Using a square knot, I hoisted the fan out the window and a couple of the other items with no problems. I had two more items to send down. Since I was running out of time I tied them together, which was not an unusual practice. The two items were heavy, steel, four foot long, pry bars. Not sure of the best way to tie and bind them together, I asked the instructor who was sitting in a chair. He wouldn't speak much less offer any advice. I did the best I could. Just a few feet after I lowered them out the window, one of them slipped and then they both dropped out of the knot. I screamed because bending below, untying the last piece of equipment I had lowered, was one of my classmates. The pry bars were like missiles locked on a collision course with his head. The instructor jumped up and watched them land on the concrete barely missing my classmate's head. They gouged out an inch of asphalt. The instructor gave me an earful loaded with expletives.

Even though I felt terrible about what happened, I concluded it wasn't totally my fault. I asked for help and got none. If you are supposed to be learning this stuff and ask a question, shouldn't it be answered? Instead, everybody just blundered around making mistakes until somebody got hurt or nearly hurt. My classmate wrote it off to good luck. My instructor gave me a zero for the day. I wrote it off to idiotic training and an asshole instructor.

Another incident that won me a break and an A for the day was when it was my turn to practice being an engineer. Prospective engineers go through special training given at the academy. Successful completion of the four week course certifies one to drive anything on wheels the fire department owns. An engineer on an engine stands outside at fires with the vehicle and operates the pumps. An aerial engineer goes in on fires except on the big burns when it is necessary to raise the 100-110 foot ladders.

Riding in the cab with the engineer and waiting while the "plug catcher" was dropped off, I hopped out when we stopped by the tower. The engineer walked me to the pump panel, (there is one on each side of the engine) and started twisting knobs and doing things.

He said, "Okay, give them water." pointing to a lever.

I grabbed the big, black, ball on the end of it and pulled down as hard as I could. I forget what pressure they pumped those old engines, maybe 120-130 pounds, but the residual pressure is much more than that when the gate is first opened. Pulling the wrong handle, the water shot straight into my chest with enough force that it picked me up and threw me about ten feet backward. One of the instructors witnessed the entire event from the third story window. He said he thought my head was impaled on the high rise inlet sticking out the side of the tower. My helmet saved my skull. My classmate who almost lost his head to me didn't feel any satisfaction when I almost lost my head to the tower. My butt and back were sore from landing and I had a headache from the clunking. I got a thirty minute time-out.

My next reaming came during an exercise when I happened upon a volunteer instructor who was lying crumpled on the second floor of the hose tower. Talking to him, he didn't respond. I was thinking, in EMS it is harmful to move an unresponsive person who is injured until he is C-collared and back boarded for protection of the spinal cord. (A C-collar and back board immobilizes the patient so in case any neck or back bones are broken the patient isn't paralyzed by unnecessary movement.) I yelled for help and two of my classmates came running. They grabbed the victim by his legs and arms, hauling him out.

I was yelling, "Wait! Wait!"

The instructor dragged me to the director of the academy ordering me to fully explain why I let other people do my work. At first he took over yelling at me and finally stopped, waiting for an explanation.

When I explained I wanted him packaged as an unconscious C-spine injury before he was moved, the captain smiled and slapped me on the back, "Okay, we have moved on from EMS and into fireground operations. Forget EMS for now and imagine you are in a burning building with a victim. Your first priority is to get them to safety. You can't worry about broken bones or anything else if flames are licking at your boots. The next time you come upon a victim, just haul them out of there the best and fastest you can. Now report back to the drill tower."

The next “victim” I encountered with three of my classmates, I knocked one of them out of the way to grab his legs and one other person grabbed his arms while we ran off with him, leaving two classmates to get lectured about stepping up to help.

Chapter 6

One morning we came to class and the instructors announced a huge fire, a third alarm, had occurred overnight. We were going to ride out there and help pick up hose. This was great news, so exciting. We would be with actual firefighters at a real scene. When we got there and all piled out, before us lay a huge warehouse with a collapsed brick wall. Hoses snaked all over the broken bricks lying in huge piles with steam venting every few feet. The smell was overpowering and my eyes began watering from the acrid air. The devastation was unbelievable. It looked like a bomb had gone off.

We stumbled through the bricks. At the butt of each section of hose, we stopped to unscrew it. Most of the firefighters were on break, sitting in chairs or standing around talking, by a big white van where Salvation Army volunteers passed out water. They were sweaty and dirty with their coats off and boots pulled down.

Fire gear back then was a little different. The boots had flexible tops to just below the knees. The tops were folded down over the hard boot. When an alarm was announced, the firefighters stepped into their boots and pulled the flexible part up to their crotch unless they were tall and then they came up to mid-thigh. After many years of crotch burns, the fire department started issuing short boots and pants with suspenders which were pulled up and worn under the coat. A few years later, after many ear and face burns, we were issued fire resistant hoods.

We were working hard when the captain of our academy approached with one of the sweaty firefighters. I heard the captain bragging about all the things they were teaching us in the academy.

"Kathy, tell him some of the hundreds of fire gases present here that are burning our eyes."

Put on the spot, my mind went blank. Never had I been asked to repeat anything like that and I was embarrassed to tell some crusty, experienced, exhausted firefighter what I had first learned a month earlier. When I looked at the devastation all around me and what these guys must have been through to put this monster fire out, gases were the furthest thing from my mind.

I finally stammered three or four gases and the captain said, "See, I bet you never knew what caused your mucous membranes to react so strongly."

They walked off together and my classmate and I went back to picking up hose.

The chaplain, in his marked car, pulled up on the scene and talked with the captain for a minute. They pulled my partner aside who was helping me unscrew butts and told him that his father had died. He sat down on the bricks, put his hands over his eyes and cried.

As I tried to comfort him, my classmate confided, "He was really my step dad but raised me since I was little. Before my step dad married my mom, we lived in poverty. He gave us love and a nice place to live."

My partner left with the chaplain who took him to his car at the academy. We didn't see him until the next week. He was quiet in school and nobody knew his dad was sick. It took the joy out of humping hose.

After we picked up the majority of hose, we were allowed to go to the van, drink diluted Gatorade and indulge in stale, plain, white bread, bologna sandwiches for lunch. Being very hungry after all that work, I chowed down despite the bland taste.

We were dismissed early that day, which was a rarity. I went to my mom's house to pick up my son. She said I looked tired. When my husband arrived, she gave him a lecture about helping me more.

"Fire school is wearing her out! Just smell her and you will know how hard she has been working."

I hadn't realized until then, how smoky I smelled. It was as if my body was permeated with nasty smoke. After arriving home, I jumped in the shower. The warm water caused the smell to waft in the moist air and fill the bathroom. Even after bathing, I could still smell the smoke on my skin, especially when I perspired. It was three days before the smoke perfume dissipated. When you walk into a firehouse in the morning, you know instantly if they had a fire the day before because that distinct and wonderful smell is everywhere.

Back at the academy, the instructors wanted us to experience the pernicious conditions of a real fire where you can't see anything you are doing or where you are going. A maze was set up in the classroom and we were given SCBAs with face pieces that were blacked out. The lights were turned off and the shades were drawn. A curtain was tacked over the door to prevent peeking. We donned our masks while on our hands and knees, opened the door, and crawled into the room. This was really creeping me out. My stomach was doing flip-flops because I couldn't see a damn thing. We went through it quickly but I was sweating like a nervous matador facing her first bull as my heart rate sky rocketed. Perhaps, the instructors, like animals, sensed our fear and toyed with us, cautioning that the maze would get more difficult after the barriers were constructed and the traps were set. My fear was taking over.

After finishing, I tried to be inconspicuous so I would not have to go through the grueling maze again. My tablemate, whom I bumped into hiding around the corner, and I got lucky, and were not volunteered for a second time that day. I stayed after class and talked with my counselor who was the same guy that wouldn't offer me any advice about tying knots. Maybe it was stupid to admit my weakness but I was desperate for some advice to shake the fear. It was hampering my ability to get through this exercise. I had come so far and it was just weeks to graduation. He offered no advice.

The next morning, I was called to the chief's office. He said my instructor told him about my anxiety.

Surprisingly, he said he knew I was an intense, imaginative person with a lot of spirit. He had seen me improve dramatically from the beginning of school and knew I could conquer my fear. The trick, he assured me, was to concentrate on the task at hand and not let fear overpower my mind. He researched my application and knew that I could speak German. He promised he would go along beside me while I was crawling through the maze and distract my fears with German speech.

True to his word, when my turn came, he crawled beside me and spoke softly in my ear in German the entire maze. One of the traps was a hook connected to a door. As we crawled by, an instructor attached the hook to our air packs. The door swung open so far and then we were stuck. We could reach back and detach it but only if we backed up to release the tension. I performed this easily but others had more difficulty from what I was told by my German speaking instructor.

"Du hast das gut gemacht!" (You did well!), he whispered conspiratorially.

We went through a tunnel where we had to take our tanks off and push them ahead of us while we bellied through. This was to illustrate how masks could still be worn with the tanks off our backs and teach us the way to squeeze through tight openings when they were too small to accommodate us and the tanks on our backs. We went around chairs, up a ramp and even had a mattress to pat down simulating a check for unconscious victims.

With the aid of my instructor, I passed with flying colors, learning a valuable lesson about controlling fear that I would use many times in my career. Focus on the task; stifle fears but let them be there in case something happens. Controlled fear will make a firefighter aware and ready for action.

We went back for more fireground operations. I became proficient at raising ladders, throwing water and all the other tasks. Two of the four women were sidelined with injuries and a few more of the guys too. With people injured, that meant more evolutions and exhaustion for those of us still standing. One of the guys who was supposed to help me raise a ladder got a hand placement wrong on his side. I raised the ladder practically by myself. We got a short break immediately after that.

I saw him hurry over to Mandy and talk animatedly, "Kathy can really pull her weight. I'll never say anything bad about her."

Mandy relayed his words, saying she was proud of me.

Later that day, after I had raised the back-breaking, 35-foot ladder several times, and did countless other tasks, we were all stopped in mid-evolution and brought over for a lecture. An instructor announced that the other woman resigned because she dropped the 35-foot ladder and couldn't raise it. All of the instructors were looking at me. The men were told to sit down. Walking me over to the beast, I was ordered to one end of the humongous ladder and one of the smaller guys was put at the other end to foot it (placing a foot on the end rung so it won't scoot). It was near the end of the day. I was upset about loosing another classmate and tired. An instructor said if I couldn't lift it and raise the fly properly, I would also be booted. In front of everyone, I grunted that ladder up, raised the fly in record time, lowered it and let the ladder drop into my free hand, then set it down while I was still five feet from the end. The instructor lurched because he thought I was dropping it but I was just getting it done as fast as I could because I was close to crying. Standing with my hands on my hips, I gave them a look that dared them to say anything. Class was dismissed early.

The truck instructor walked me out to my car where I cried. Why did she go? I wanted to know. We would be graduating next week. He told me she was informed she could be fired or resign. Resigning would be better because she could work on her strength, reapply and maybe be hired again. If she was fired, she would be barred from applying ever again. All recruits who are thrown out of school are given the same option. Most everyone resigns. They were going to have to fire me if they wanted to get rid of me. Never, ever would I resign. My instructor said he already figured that out. For the rest of the fireground evolutions, I was the only female out there.

I tried calling the woman who resigned but it was many months before she returned my call. She said she had another job and was happy. She was grateful to the fire department for all the things they have done for her son. He has muscular dystrophy and the fire department has a camp for kids with M.D. Her son went every year. As for the instructors, she said they could all go to hell and they weren't representative of the good on the department. As far as I know, she didn't apply again. Who could blame her.

Even though I finished the fireground evolutions as the lone female, I felt fortunate to have had women in class with me. In later classes, many times there was only one woman, some times only two. In one class there were no women and only a couple of African-American males. In the picayune atmosphere of the academy, it had to be torture to be one of a kind. I had a putrid, albeit small taste of what it was like to be the lone woman. For most of my career, I have been one of a kind in the companies. Fortunately, as my brother had promised, it was agreeable and satisfying, even fun like a slumber party, after the boot camp attitudes in the academy.

Our class still had one final test to take before being released into the companies as probationary firefighters, the live burn. We were reporting the next day to a house donated to the city to be burned for the purpose of training the recruits and testing what we could really do with fire and water. Had the training paid off?

Chapter 7

The house to be burned was a two story, residence, frame construction, on a large chunk of land. It had a big, uncovered, back porch. We put up ladders, climbed on the roof and swung our axes into real shingles and wood, chopping holes in the roof. After poking a multitude of holes up top, we went through the house in mock search and rescue. Then came the coup de grace, the fire to extinguish. When it was my team's turn to go in, four to a team, I was second in line.

The instructor said, pointing at me, "No, we want you on the nozzle at the front of the line."

I assume, because I suffered from claustrophobia, they wanted to see if I would waffle when it came to the real thing. We switched places on the line, kneeled on the porch, and waited for the signal to enter. We were wearing all of our gear, including our masks. Our tanks were turned on and we were breathing bottled air.

It is strange to breathe through the masks. It is no problem taking a breath; plenty of cool air comes in. It feels like air conditioning on your face, especially when you are sizzling hot from the gear. People feel sorry for us in the winter when we have icicles hanging from our helmets, and our coats are heavily frosted with ice making it nearly impossible to unhook the buckles. Our feet and hands might get a little cold, but as long as we keep working, we stay warm. Summertime firefighting is when people should feel sorry for us. We work extremely hard in cumbersome gear, carrying heavy equipment. Dragging an unwieldy hoseline, trudging into a roasting fire, we put it out and exit overheated. Getting out of that gear quick enough and cooling off fast enough doesn't happen. Summertime is heart attack weather for firefighters.

Back to the breathing, it is difficult to breathe out in the SCBA. Normally, breathing out is a passive task. Your muscles relax and the lungs equalize with the surrounding air pressure. The mask alters that with a valve that must open and shut making it necessary to force air out resulting in a conscious effort.

Waiting, hearing my own rapid, Darth Vader-like, breathing, the anticipation was killing me, making me a nervous wreck! I had never done this and wasn't sure if it would be just another step in training or a jump off a cliff! Finally, we got the signal to go. We crawled through a front room to the right. We didn't have to stay on the wall with a hand since we had a hose line to follow out.

The instructors were always trying to trick us into mistakes and one of the mistakes was to stand up and walk. In fires, floors can give way due to the fire or be booby trapped by arsonists or be structurally unsound. A firefighter's weight, spread over a large area, is less likely to cause a fall through a floor. We are unlikely to crawl into a hole but it is probable we will step into one. Well trained, we crawled through the room. As soon as I turned the corner, I saw the flames. In the opposite corner of the room was a pile of wood with a roaring fire. There was some smoke but it was up at the ceiling and hanging around the instructors' heads who were stationed on each side of the room, observing. Very clearly, I saw the orange and yellow color of the fire and the smoke curling upward from it.

Immediately, before the rest of the team had rounded the corner, I opened the nozzle and poured a flood of streaming water at the base of the fire, just like we were taught, taking only a few seconds to extinguish it. I'm not sure the rest of the team even got to see any flames. The instructors shouted for me to shut the water off which I did right away. I was thinking this wasn't so bad. In fact, it was really kind of easy and I was feeling victorious. Veni, vidi, vici. (I came. I saw. I conquered.) The instructor, over by the fire, started yelling at me about a technicality. He wanted me to come all the way into the room before I opened the nozzle. I deflated quickly. We crawled back outside, dragging the hose behind us, and took off our hot gear.

The other instructor followed me out. He said he did not understand why the first instructor was upset. He conjectured that the first instructor was hoping I would have more trouble. He said I did an excellent job putting the fire out quickly and felt I would do okay in the companies. Oh my god! An endorsement from one of the instructors! I must be dreaming! It couldn't be real! It was just a small thing. I would do okay in the companies. I had put out the fire quickly. But, after thirteen weeks of hell, it seemed huge! Compliments were seldom given to anyone but incredulously, I got two!

By the end of the day though, I was seething. All day, I waited for the men to have to prove they could raise the 35-foot ladder but it didn't happen. I had been singled out because of my gender, my competence questioned because I am female. What kind of message did that send to my classmates? It was an issue I would face again and again after the academy.

Now that we had slain our worst fears in the house, we were given the fire department's version of a dream sheet. In the army, I wrote my first three choices for where in the world I wanted to be assigned. I never knew anyone who got their dream pick. There were only three places to choose on the fire department, A, B, or C shift. Indianapolis firefighters work a 24 hour shift followed by 48 hours off, one full day in every three. The A, B, C, designation refers to the order of the shifts. The A shift works a 24 hour period and is relieved by the B shift which works 24 hours and is relieved by the C shift. The A shift is back on after the C shift and 48 hours off, beginning a new cycle.

My brother was on the A shift. I already knew about trading time, which was having a person, any firefighter, work a shift for me and then I pay them back when they need time off. Pay back is hell because a firefighter has to work 48 hours straight to pay back a traded day. But trading just one duty day results in getting five calendar days off in a row. Trading two days nets eight days off. Firefighters on IFD can trade up to three days and go on vacation if they want.

In my first ten years on the job, the only way I could get time off in the summer was to trade time. Vacation picks are based on seniority and the guys with 20 plus years take all the summer vacation. We used to have to take our vacations in two to four week increments. I think around my seventh year it was changed so we could take our vacation days any way we wanted. Even now, with 20 years on, I still can't get much summer vacation and it is hard to get a weekend off. It is a good thing we can trade time or the young guys would never get a vacation with their kids.

I put my first choice as C shift, second B shift, and last A shift. I thought my brother and I could trade time all the time if we were on different shifts. I chose C shift over B shift. Due to a quirk in the calendar, leap year, B shift works more holidays than the other two shifts. I wanted to be off for as many holidays with my family as possible. Some like the B shift because of the holiday pay that is presently $100. We turned in our dream sheets and waited to be told what shift we would be assigned.

On our second to last day, we were given final lectures by our instructors. The P.T. instructor asked us to faithfully continue our exercise program in the companies.

He said, "You will be glad twenty years down the road. Very few people in the companies work out even though there is a mandatory fitness program. The old guys are in bad shape. They are overweight with poor cardiovascular fitness. I can't imagine how they don't drop dead from heart attacks."

"Be sure to wear your SCBA's on every fire," pleaded the upper respiratory afflicted captain.

The assistant chief of training, whom we saw for hazardous materials class and little else, gave us a lecture about what it would be like out in the companies. (He looked like one of the fat guys who never exercised that our P.T. instructor talked about. He already had one heart attack and had to retire after his second myocardial infarction a few years later.) He said we shouldn't question any orders we were given. If we were ordered to go buy a case of beer, do as we were told. And we would be asked to buy beer. We should be an hour early for duty, the first to answer the phone, the first to jump up after eating to wash dishes, the first to start housework in the morning, always volunteer to take night watch-the person who slept in the radio room and hit the okay button letting control know we were aware of the run. There was other advice about how to take hazing because everybody would be tested.

Another instructor told us not to be 23-hour-firefighters. Everybody hates a 23-hour-firefighter. We were wondering what a 23-hour-firefighter was. Of course, I asked the question. It is one who is relieved at 0700 but does not report until 0755 (Our reporting time now is 0800. Don't ask me why it used to be 0755). It was easy to cure somebody like that fifteen years earlier because there was only A and B shifts. If somebody relieved late, she could turn around and relieve him late. That would either cure the late-reliever or they would continue the late-relief cycle so it was impossible to have a 23-hour-firefighter. With the inception of three shifts, two people had to work together to pay back the offender, which did not happen. Therefore, the 23-hour-firefighter was born.

We were warned that word gets around quickly if one is a bad substitute. When reassigned to another house, the regular guys will call ahead delineating the sub's poor qualities over the Centrex-the fire department business phone that has to be answered within two rings announcing the station number, rank, and name, followed by sir. (I never say sir until I know the gender of the speaker and then I will say sir or ma'am.) The next company knows before the sub reports that he or she is a bad substitute and needs to be taught a lesson. At the next station an incompetent sub would be given a cold reception, extra work and receive an ass-chewing about every little thing. A really bad sub would be reported to the chief who would send the sub away, not assigning him to that house anymore, moving him or her from station to station on a daily basis.

The lecture continued. To be a good sub, a firefighter should be quiet and join in whatever activities the others were doing. For instance, if there were four guys who played cards together in the evenings and one of the regulars was gone, then we should jump in and play cards. A sub should be sure to bring change, in case the card game is poker. If it was volleyball or basketball being played, we should join in if they needed us, even if we couldn't play very well. We should at least try and we would be appreciated for our efforts. Most importantly, we should never miss a run and always do as instructed by the officer on any EMS or fire run. They reminded us that we would probably be the ones to take blood pressures and heart rates since most of the guys weren't happy with ambulance work. We should jump right in with both hands on the medical runs without being told or asked; get out the equipment and start working.

Finally, our shift assignments were revealed. Just like the army, I got my last choice, A shift. So much for trading time with my brother. As time went by, I realized it was better that I was assigned to A shift because it is hard to have family functions when two days out of every three are unavailable for celebrations. It is easy to find somebody to trade time with especially if one is reliable about returning the time. The rule is, a firefighter traded time when you needed it so you should make every effort to repay the borrowed time when he needs it. By getting A shift, even though I felt slighted at the time, they did me a favor. The instructor, who was mad because I put the fire out so fast, said the C shift chief wanted the best people on his shift, so they put Mandy there. Also they needed women on the A shift because there were none, so Mary and I went to the A shift. Mary stayed on the A shift for 20 years. Mandy and I spent time on all three shifts.

We reported one more time to the fireground for fun competitions. After we raced up some ladders, all of the A shift people were ordered to report to an instructor. With him was a tall, African-American male with starched, dress pants and a white shirt. His shoes were spit-shined and reflected up. He was sharp looking, clean shaven, getting some gray at the temples in his neatly, trimmed hair. His gold badge reflected the sun as he moved.

He was introduced to us as our new chief on the A shift. I was impressed. The cadre at the training academy wore blue shirts. I hadn't seen many people in white shirts and later found that all chiefs wear white shirts, whether they are downtown or in the companies. He welcomed us to the A shift telling us he wanted his subs sharp looking with shoes spit-shined, uniforms pressed, moustaches trimmed, hair cut and neatly combed. (He was a stellar example of what he expected.) He ordered us to report at least an hour early to our companies on Saturday morning and every duty day thereafter no later than 0700. He said he respected the training we were completing and knew we would perform to expectation.

He dismissed all of the men and asked me to stay a minute. He told me the men in his companies would be friendly and helpful. He wanted to know immediately if I had any trouble with anyone. He also warned me to watch out that none of the men got too friendly because he was sure that would happen and I should be on my guard.

"Yes, sir, Chief."

"All right, you are dismissed and remember, these men are like dogs in heat. Congratulations!"

He was a nice person and in the years I was under his command, he treated me with dignity and respect, even protecting me when I made mistakes.

As I returned to the fireground competition, an aerial and an engine pulled into the lot. The crews got out and approached me.

"You are coming to our firehouse on Saturday," they informed me.

I hadn't been told yet where I was going; only that it would be A shift.

The men were all talking to me at once, crowding around me, "Yes, you are reporting to our house. We've already been told. We came to meet you and welcome you. Can you make cookies? We'd love to have some home-made peanut butter cookies."

They introduced themselves. They all had nicknames. The one who asked me to make cookies was nicknamed "Cookie". You can probably guess why.

We had just received our lecture on being polished and clean cut. I couldn't believe how these guys standing in front of me looked! Their shoes were brown; it had been so long since they were polished. Their hair was scraggly. Their moustaches drooped below their chins. Their pants were wrinkly and the butts were light blue from wear. Their shirts were paper thin with frayed collars and a hole here or there. They smelled okay and had shaved but they were about as far from spit and polish as you could get. I promised to make them cookies and we said good-bye.

In my early years on the job, the department gave us a yearly allowance for uniform replacement. Inspections were sprung on us and we had to present ourselves. The older guys were wise to this and saved a set of clothes that they never wore so they looked spiffy for inspection time. They spent their clothing allowance as vacation money. After I got the hang of it, I did the same thing. Today we have a quartermaster that issues all of our clothing but we still receive a partial allowance.

When I went back to my classmates, I asked if they got a gander at those guys. We were getting a glimpse of what life would be like from now on and we couldn't wait. We were cut loose early and went for a drink at a pub close by. I had a couple of beers and was feeling giddy, laughing like crazy. We were all toasting each other, sitting on laps: boys sitting on the instructors' laps, boys sitting on boys' laps, girls sitting on girls' laps, girls sitting on boys' laps, girls sitting on instructors' laps. We were ecstatic to be finished with school and moving on to the companies. Tomorrow we would graduate.

I left before I drank anymore. I'm a lightweight when it comes to booze, three drinks and I'm dancing on the table, four drinks and the room starts spinning, five drinks and I'm in the toilet puking, six drinks and I've passed out. I didn't want to become an impaired driver, get pulled over and be fired before I made it to my first duty station. I drove home and slept peacefully, like I hadn't been able to for months. I wasn't dreaming about fire triangle/tetrahedrons and tossing and turning over ladder raising techniques. The beer helped soothe too.

On our final day, I informed the chief of the academy that I was changing my last name to my family name, "Gillette". Always interested in gossip, he asked me if I was getting divorced. I said I had been married at 18 and divorced two years later but had never changed my name back. When I remarried I didn't change my name. Now that I was starting a new career in the companies, I wanted to be known by my brother's name, our family name, so everyone would know we are related. I was issued a new name tag.

When I told Mandy, she said, "Gillette. Firefighterette Gillette. I like that."

We practiced our graduation ceremonies that day. Class rank was announced. I graduated seventh in my class. If I hadn't had that bad week when my niece died, I would have been higher but I was happy. Nothing would bring me down. The instructors said they were going to break precedent and award top recruit to the number two ranked student because of his help getting the class through the chemical part of hazardous materials. I'm sure number one was disappointed but he didn't show it. Class ranking is important because for the rest of our careers we would be called in that order for vacations and putting in for regular positions. Sometimes, only a portion of a class could go regular (be assigned a permanent position in a company) and the lower ranked ones might have to sub for another year or more. I was glad I finished in the upper ranks. The hard work and studying paid off in future years.

We were issued our helmets and badges. We pinned the badges on each other and tried our helmets on for size. Mine seemed slightly big but it was adjustable and I finally got it to fit okay. It was still kind of wobbly but I got it to stay put by tightening the chin strap. We were released early but had to be back with our families at 1900 hours or 7 PM. We would be using 24 hour time since we had 24 hour shifts. We were given a brief lesson to refresh us on how to tell time by the 24 hour clock and headed home, excited to be returning in the evening.

My entire family came to my graduation ceremony. There were awards. The number two graduate got a watch for outstanding student. A mayor's deputy gave a speech to the class. The chief of the fire department congratulated us and our families. All of our instructors had a little something to say. We presented them with a plaque from our class, the 57th recruit class or as we called ourselves, "Heinz 57", because we are saucy. We had secretly chipped in two weeks earlier when our pictures were taken. The plaque hangs in headquarters with our class picture. There are many plaques from other classes. Whenever I am detailed to headquarters, I look at our plaque and picture, thinking fondly of my classmates, but am glad those days are over.

My husband and I stayed for a few drinks and I had a piece of cake, then two pieces, at our graduation party with my classmates, meeting some of the spouses. One third of us had to report the next day, Saturday, our first day in the companies. B shift reported Sunday, and C shift Monday. I didn't stay late. I wanted to be at the firehouse at 0630 and I wanted to be fresh. Leaving with another piece of cake for me to eat in the car, we drove home. As soon as I entered the door, I went to bed.

Part Two

The Real Fire Department

Chapter 8

I was so excited I had a tough time sleeping. I woke at 0530, a half hour before the alarm. I knew the traffic would be light since it was a Saturday. Killing time before I left, I went into my son's room and looked at him sleeping. It was difficult not to pick him up and swing him around. I was overjoyed. He would be laughing with me. He was such a happy boy, finally. He was cranky the first year of his life and didn't sleep through the night. When he finally did sleep through the night, I woke with a start, thinking he died. I didn't want him to stop sleeping through the night now, just because I couldn't keep my hands off him. I loved him so much. Forcing myself to leave the room, I quietly shut his door.

Messing around in the kitchen, rearranging silverware in the drawers, wiping counters, dusting, I could stand it no longer. I left at 0605 for the firehouse, arriving there at 0625, making only one wrong turn. It was a large, single story house with four bays, sitting back off the road. Driving down the side road, I saw the entire back parking lot was fenced with the gate secured by a huge padlock. I went around the front and looked in the windows. Not seeing anybody moving around, I didn't dare ring the doorbell and wake anybody. The last thing I wanted was them calling on the Centrex and telling the next company how I woke everybody. I went back to my car and waited 30 minutes until the first person arrived. Before he got the gate unlocked, there was a line of cars waiting.

It was 0705 and I was hauling my gear into the building. I was told to set it by the door in the day room, very close to where I entered. One of the guys helped me bring some stuff in. They taunted him, "Why don't you help me put my stuff on the engine? What makes her so special? You hardly put your own stuff on the truck?"

The Captain told me to come sit at the table and have a cup of coffee with him. I don't drink much coffee because I don't need the caffeine. I am overloaded with energy. The words of the instructors telling us to join in came back to me so I got a cup, sugared and creamed it and sat at the end of the table where I could jump up to refill coffee cups. The shift getting off was still there and most of the A shift had arrived. It was a long table but we were still squeezed together. Nineteen guys were there, and me, ten on each shift. Thirteen of them were sitting at the table, stealing glances at me. The others were standing close by. The talk was loud, punctuated with laughter. These guys knew each other well and enjoyed one another's company. I sat listening. I didn't know what to say since I wasn't acquainted with anybody. Cookie asked me if I brought the stuff to make cookies. I nodded yes.

In the background, the radio was going off every once in a while but nobody seemed to pay much attention. About 0730, one of the men looked at me and said I would be on the truck today behind the officer. I got up to put my gear by the truck. This was a newer truck with a partially enclosed cab. It seemed big and long compared to the little, old engine we hung off at the academy. I would sit backward, a whole new experience. There was gear in my seat and hanging from the bar. The officer told me it was the C shift's and he hung it up on the truck rack. I got my gear organized in the order I would be putting it on, boots on the floor, gloves on the seat, helmet next, leather tool belt, and coat on top. I hooked my face piece to the mask. The officer asked if we were all issued our own face pieces. I said just a few of us because the regular face pieces were too big so we were fitted with smalls. I turned the valve and checked the tank to make sure it had plenty of air.

One of the guys showed me to a locker in the men's locker room, the only locker room. I shoved my stuff into the locker, grabbed my shorts and t-shirt and went to the public restroom up front where I changed. I came back with my civilian clothes, knocked on the locker room door and was told to come in. I sat down on the bench and was busy organizing my locker when a guy came in and got completely naked next to me. His backside was showing and I couldn't help but look since he was standing incredibly near.

A guy on the opposite side of the corner noticed him too and said, "Ahhh, Close, do you always have to show your butt. Give her a break."

"I'm not changing the way I do things for her or anybody and they can kiss my ass if they don't like it."

I figured I could arrange my locker later when everybody wasn't busy changing. I got out of there and picked up a broom to start sweeping the kitchen. The captain saw me and said if I planned to shower that I could make an announcement over the P.A. He showed me how to work the phones and the P.A. which was something we hadn't learned in school. After sweeping the kitchen and mopping it, I went out to help in the bays where they were washing the apparatuses. This house had a squad, truck, engine, and an apparatus we called a "Squirt". It had extra hose and a nozzle that could be extended high in the air for defensive firefighting. The Squirt had square wheels which meant it didn't go on many runs. IFD no longer uses this equipment.

As soon as I turned the corner, one of the guys had a fire extinguisher in his hands and sprayed me with it. The big, 2.5 gallon, silver cans are pressurized and can shoot water 20 feet. I tried running the other way but he chased me. I turned on him, and backed him into a corner by the bay doors, wrestling him for the can. He let go and ran off.

I turned ready to squirt the person coming behind me and realized why I had been able to wrestle away the extinguisher so easily.

"Uhhh, Chief, sir, good morning, sir," I stammered as I stood with my hair dripping and socks sagging, ready for a wet t-shirt contest.

The chief just shook his head and sarcastically said, "Welcome to the fire department, Private Gillette."

Finding the hose to pressurize the extinguisher, I filled it and put it back on the engine. I grabbed dry clothes, changed and went for a run with a department radio in my hand. The outside speakers were on but I didn't want to take any chances missing an alert so I held the radio close to my ear and listened to the station speakers. After running a quick pace for 30 minutes, I went to lift weights in the little, workout area behind the engine. While I was bench pressing, the alert sounded and some guys ran for the apparatus. Dropping the weights, I jumped up and ran for my aerial. They stopped and started laughing.

Returning to my workout, I hopped on the stationary bike to cool down. Another alert sounded. Again, I ran with them. They stopped and laughed some more.

Another alert and everybody started running, "Come on Gillette, this is the real thing. It's a hospital."

Double timing to the truck, I hopped in my boots, grabbed my coat and jumped on. The truck pulled out behind the engine, turned east and before I could get on all of my gear, we were there. It was a hospital just across the river.

Quickly cinching my SCBA, I jumped down and started walking toward the building when the officer stopped me and said, "You forgot your gloves."

I hopped back on and grabbed my gloves.

When I started to walk away, he said, "Don't forget your axe."

I hopped back up and grabbed the axe.

He looked at me and said, "Okay, now we're ready."

As we walked toward the doors, I noticed I was the only one wearing an air tank. A voice said something over the radio and everyone did an about-face. I wanted to know what was going on. The officer said we were disregarded. I looked, blinking at him, and he said it was a false alarm, no fire, so we were disregarded, and could go home. He explained that almost all of the hospital dispatches were false alarms so I shouldn't get too excited about any hospital runs.

When we got back to the fire house, I got out of my gear and went back to the bike to finish cooling down. While I was closing my work out with stretching, a couple of the guys came over to me, including Mr. Naked Butt, and handed me a piece of paper. They said it was a printout of the run we just had and they thought I might like to save it.

"Sure. Of course. Thank you very much!"

I thought that was extremely sweet. These guys were surprisingly nice and thoughtful. Unfortunately, although I had it for many years, I lost that run sheet somewhere in my travels along with various other items.

It was getting close to 1100 and workout time was from 0800 until 1100 in those days. Making a loud announcement on the P.A. that I would be in the locker room, I showered, shampooed, dried off, and dressed in about five minutes. It was slightly unnerving to be in that huge locker room by myself, and trusting that the P.A. announcement would keep everyone out, but it worked. As soon as I finished, I announced the locker room was now open. I didn't want to be responsible for anyone busting their bladder on my behalf.

I went to help the cook, setting the table. Anytime the phone rang, I jumped to answer it.

A couple of times, the calls were for guys on the other shift and the cook told me, "They'll be here tomorrow." or "They were here yesterday." so I relayed that.

When lunch was served, I was starving and ate as much as any of the men. Busy eating and listening, I did not say a peep. The place settled down after lunch. Some of the guys were playing pool or watching TV. Some were napping on the couch or in a chair. I was pacing from room to room, too nervous to sit down. Every time the alert would go off, I walked toward the bay because I was not sure when my station would be called. The radio went off a lot in those days. The chatter was incessant. During the day we heard all the runs in the old city limits. Box runs were being dispatched with numbers like 145 or 146. Each house had a list of thirty or forty box runs.

Those runs were for multiple residence houses or businesses where a lot of companies responded to call-ins or alarms. The hospitals due to their large size were box alarms. That was why I could be easily tricked. Anytime control gave out a number over the radio, I had no idea if it was my aerial being dispatched on a box run. Later, a nice guy, feeling sorry for me, pointed out documents above all of the doors, which listed the box runs to which our house responded. No longer could I be tricked.

A couple of times that day, the squad was dispatched on EMS runs and I watched them leave. My truck officer was sleeping on the couch during all this time and never stopped snoring. I was thinking he is going to miss a run if we get one, or maybe someone will wake him as they pass. I was standing next to him, watching the guys play billiards, when the alert sounded. As I saw everybody moving toward the bays where the apparatuses were parked, I turned to give the officer a shake. He was already awake and sitting up, slipping into his shoes. I didn't see anybody wake him. How did he do that? He said this was a possible fire, so I should be ready and just stick with him, follow him around like I was a puppy at his heels.

The thrill of going down the street with lights flashing and sirens blaring gave me goose bumps. When we got there, which seemed like instantly, I had all my gear with me, since I had practiced getting it all on the first run, working out the jitters. I jumped down off the truck with my air tank wobbling and clinking on my back. Already, I was hot in the afternoon sun and started to sweat in all the heavy, black gear. I was standing behind my officer when he was approached by another red helmet. (Privates wear black helmets, lieutenants and captains wear red helmets and chiefs wear white helmets.)

After conferring, he turned around and almost ran into me. He laughed heartily and said I didn't have to stand that close to him. We were disregarded. It was another false alarm.

Disappointed, I was anxious to get a fire under my belt. Still, I was relieved it wasn't a fire. My adrenalin rush was subsiding for the second time that first day. I could feel my legs and arms get a little wobbly walking back to the truck. It felt good to take off my gear and have a breeze blow over my skin. We hopped back on the truck and pleasure drove back to the station.

When we returned, everyone helped set up the volleyball net. They asked me if I knew how to play. The person I replaced normally played and they needed me. I love volleyball and played throughout high school and the army. I was ready and happy to play, seizing a chance to work off some of my pent up energy. We all got out there, playing jungle volleyball. There was no setting and they only spiked the ball when one was hit accidentally close to the net; get the ball over as soon as possible was the idea.

There were a few rules, like you had to stay on your side of the net and the ball had to stay in bounds but just about anything else went. All the carries and lifts were okay. There wasn't much diving or digging for balls since we were playing on an asphalt parking lot but there were a lot of toes trampled with a few people falling down and scraping knees or elbows. I went down one time when a low returned ball was spiked right into my face. I was fine except for being totally embarrassed that I missed the ball. Poor Mr. Naked Butt got ribbed for spiking the ball into the little sub's face.

We played for about an hour when the alert went off. Two of the guys on the EMS squad had to run off for a medical emergency. We continued with eight people until the squad got back about 30 minutes later and rejoined us. Two of the other guys left then to start cooking. I offered to help but they said I should stay so the sides would be even for the game. We played until the cook said we had 15 minutes before dinner was ready. Everyone else showered while I washed my hands and red, sweaty face, then helped set the table.

The meal was delicious and everybody complimented the cook. They wanted to know where the dessert was. Oh my god, I forgot the cookies! I piped in that I would make desert right after the dishes were done. As soon as the floor dried after being mopped, I whipped up a batch of peanut butter cookies using a recipe from a cookbook in the kitchen drawer, as I had promised. They taste so good when they are hot out of the oven. Most of them disappeared right away. They ordered chocolate chip cookies for the next day.

The officer wanted to know if I had made my bed yet. I got my garbage bag full of sheets, blankets, and a pillow, followed him to a corner of the bedroom and made the bed he said was mine. He told me I would need to make it in the mornings so I wouldn't disturb anyone throughout the day by wandering in the bedroom to make my bed later in the day. The room was huge, filled with 12 single beds. It was an open bay with air conditioners blasting. It felt like winter in there and I was glad I had brought a blanket and a comforter.

Runs were announced over the radio until 2300 or 11 pm when it was switched from day watch to night watch. Some of the guys had already gone to bed at 2100 and 2200 hours. It was astounding they could sleep with that radio blaring on and off so randomly. I watched TV for a short while and then brushed my teeth. Slipping quietly into the bedroom where a couple of the guys were snoring, I laid there and heard a fart or two or three or four. Finally, after turning over about twenty times, I farted and fell asleep. It seemed like right after I got to sleep, the lights came on, the radio blasted something, and the two guys on the squad got up for an EMS run. Everybody else just rolled over.

Getting out of bed, I walked out to the bay to watch them leave. I paced around, watched TV and shot a few balls on the pool table until I heard the bay door open and the beeping of the back up alarm as they backed expertly into their parking spot. I tip-toed back to bed but tossed around a lot before I could fall back to sleep. At 0300 the squad got another run but I didn't get up with them. I was too tired. I rolled over like everybody else but I woke when they came back, hearing the backup-beeper. Around 0600, I heard some of the guys get up. I had to pee so I got up too. After seeking advice about the number of scoops, I made a pot of coffee. Most of us were in our uniform shorts and t-shirts in which we had slept but some of the guys were already showered and in civilian clothes. By 0700 most everybody was sitting at the table talking. The other shift started coming in and I knew the people from my class who had been assigned to the B shift were probably already sitting nervously at a fire-house pouring and drinking coffee as I had but I was going home with my first day successfully completed.

Informed that I would not move to another station for a couple of weeks, I could leave all my gear. The men were talking about the new class of recruits infiltrating the companies. Some of the B shifters introduced themselves to me. Around 0730, they said I was relieved and could go home. Wandering out to the bay, I took a last look at the truck. My gear had been removed and put on a hook with my boots under my jacket. Noticing that my face piece wasn't there, I took it off the SCBA and put it in its bag, hanging it from my coat. Standing there, staring at the truck, engine, squad and squirt, I could hardly believe I had made it through fire school and was now on the real fire department. Even though I had barely slept the past two days, I didn't feel tired at all. I felt happy and proud to be standing in that spot, unable to take my eyes off the equipment. The momentous first day and a sense of accomplishment made me high.

On the way home in the car, I started feeling sleepy but when I pulled into the driveway, I bounced out of the car and ran into the house to see my baby. He was awake in his crib waiting for me. He stretched out his arms. Picking him up, I swung him around. His curly blonde hair was getting long and his cheeks were chubby. He was like an angel. About 1400 in the afternoon, we both were tired and laid down for a nap together. After nursing him to sleep, I woke up just in time to fix dinner. Waking my snoozing husband on the couch, we ate spaghetti dinner, which Alex slurped all over himself. I was pooped and fell asleep on the couch. I returned to consciousness every once in a while and heard the boys doing stuff, but I was pretty much zombied.

My husband woke me to kiss the baby good-night. I shuffled into the bedroom without brushing my teeth, falling immediately back to sleep. The baby woke us about 0700 and it was nice that I didn't have to work Monday morning. My husband had to be at school at 0900. I fixed breakfast and lounged around with my son until lunch when we went over to Mom's. She said I looked perky and rested. Fortunately, she hadn't seen me the day before when I had bags under my eyes because I would have been lectured about working too hard.

My brother came over and we all went out to lunch, laughing about my first day in the firehouse, especially when the stuffy chief saw me dripping wet. Telling my brother that no one said a cuss word the entire 24 hours, I had to swear I was telling the truth. He said that didn't seem normal.

Chapter 9

The next day I showed up for work at 0700 so I would not be sitting outside the gate waiting again. My timing was perfect. I rolled through the gates, jumped out of my car, ran into the firehouse, and grabbed my shorts and t-shirt out of my locker before Mr. Naked Butt got there. After changing into my uniform in the restroom up front, I went to the kitchen to pour coffee for everybody. They told me I was on the truck again, so I went to the bay to put my gear on it. I didn't know where to stash the stuff I was taking off the truck but was directed to the C shift area where the racks were mostly empty. Hanging up his equipment, I sighed with satisfaction, the first time I had put away anyone's gear.

The day was pretty much the same, the squad rolling out for EMS a lot and the engine and truck following the squad on fire runs. With enough time to get an extensive work out done and reduce my energy level, I settled down a notch or two. After a few workdays on the truck, the regular person came back and I moved to the engine. Mr. Naked Butt came to me the first day I was on the engine after I finished my workout and showered. He said the officer wanted me to show him a sample of what I learned in the academy so they could judge if they were teaching us anything worthwhile. I demonstrated my skills of hooking up to a hydrant, turning it on and throwing water with the 1½" line. There is a hydrant at the end of the approach of the firehouse so we did all this while we were in service waiting for a run. Of course, I had no problems and they all agreed that maybe the academy was doing something right.

My first fire came the next day. It wasn't a blazing inferno and I didn't even put on my SCBA. It was a big, municipally owned, plastic, trash dumpster that was half melted by the time we got there. The contents were spilling out of the melted side and I got close to the edge of it, peering inside to get the water directly on the fire. We were stepping all over the trash, walking around, attempting to spray it from every angle. One guy got a long pole, a pike pole, and was raking the trash around while we sprayed water over the stinking mess. As I moved, my boot got hooked on the edge of a dirty, old, piece of carpet someone had thrown away, causing me to fall backwards on my butt, landing on a tin can. The pain shot down to my toes. With difficulty, I slowly got halfway up. The guys shut off the water and helped me straighten up. The officer asked if I was okay. Shaking my head yes, I walked bowlegged back to the engine.

My butt hurt for four weeks after that. I thought about going to the doctor to make sure I hadn't done something terrible like break my tailbone, but time heals all. Hating to admit that I had my first injury before I moved to the next company, I ignored it and it eventually went away.

We came back from that run arriving to smell the delicious food and sat down to a tasty, hot meal. Sitting only on my right cheek until it was about to go dead from the wooden bench, I switched to my left cheek, carefully avoiding the painful middle.

The guys talked about the huge, plastic dumpster melting causing the poor sub to trip and "when would the god damn city learn that plastic garbage cans won't work in the ghettos."

Everyone stopped chewing and looked at me. I looked at them wondering what I had done wrong.

The officer finally said, "I hope we haven't offended you? We didn't mean to."

I just stared because I had no idea what he was talking about. Explaining, he said the chief had given them a lecture before I came that there was to be no cussing or anything else while I was there, and they had better be on their best behavior, or they would all be downtown in the Chief's office explaining their bad manners. Not sure how to interpret my disquieting silence, the captain expressed his hope that I wouldn't report the errant behavior to the chief. No wonder my brother was puzzled when I told him how clean their language was!

I declared to an attentive audience, "Well, he's right." I paused, letting their hearts skip a beat; "You should treat me like a fucking lady."

Their eyes bugged open and the food dropped out of their mouths as they laughed. I thought they were going to roll backwards off the bench seat onto the floor. They were beating the table with their silverware and punching the guy in the arm who blurted out the cussword. I think they were all relieved they could go back to normal. The person next to me put his arm around my shoulder and squeezed me telling me I was going to fit in just fine.

We played volleyball almost every day even though my butt ached from the jumping around. One of those days, while the officer was off, I had my first garage fire. Naked Butt, catching the seat, or riding in the officer's place, wasn't used to the duties of lieutenant. The other person on the back step hopped off to hook the hose, a supply line, to the hydrant. Almost all rookies, especially when they are brand new, stay on the engine with the officer, or the acting officer in this case, so they can gain experience being on the nozzle and spraying water on the fire. Mr. Naked Butt, the substitute officer, helped me pull the hose off the engine. Our engine and ladder were the first ones there so we were in a hurry to get water on the garage to prevent the fire from spreading, what we call attacking the fire. Bracing myself, as I was taught in the academy, for the first blast of water, we opened the line and he pointed where I should direct the water shooting out of the nozzle.

A bunch of crackling and shouting came over the radio. He rushed to the back of the engine to help with the problem. He came back after a few minutes, a little flustered. The radio traffic was the engineer, driver of the engine, shouting that the hose was not clamped from the hydrant and the hydrant had to be shut off before he could hook up. My sub officer had forgotten to put on the clamp. It's easy to do when in a hurry and it is not the normal routine. Several of us had forgotten the clamp when we played officer in fire school but here was a guy spacing it with a lot of years on. They were asking him what was wrong. Maybe he couldn't keep his mind off the new firefighter and on his work? Ignoring their teasing, he told me I was doing a good job putting out the fire.

Other companies arrived and set up hose operations from the other side. My sub officer took the line from me and told me I could take off my SCBA since the fire was dissipating. I noticed I was the only one with a tank on my back. The engineer helped me shed my breathing apparatus and I ran bowlegged, since my butt hurt, back to my hose line, grabbing it greedily. The crew on the other side of the garage was spraying water at the top of the roof causing chunks of boards to fly off in our direction. One of the boards clunked me on my helmet with a loud thunk. My acting officer told me to stand my ground and he would be right back. The crew on the other side, after he did some shouting, shut down their line and let us finish the attack.

After the fire appeared to be out, we shut down our line and let the truck crew enter the remains of the garage to chop the wooden walls, checking for hot spots, until we were sure the fire was totally out so there would be no rekindle. We sprayed the garage down once more, soaking it, because it is a serious embarrassment to a company if they have to come back to the same fire later on. An officer gets a bad reputation if they have that happen a couple of times.

Everybody except us picked up their hoses in anticipation of being released from the fire scene. The truck crew shuffled around in the garage a while longer while we helped load all the hose back on the other engine. We doused it once more for good measure, and with the help of the truck crew, picked up our hose, then headed back to the station.

Having been at this firehouse now for a few weeks, I habitually announced my shower time. The men were very respectful and there were no problems except for one minor glitch. I was in the shower when a run for the engine came in. Afraid they might leave without me, I didn't bother to towel off. I threw on my shorts and t-shirt and ran out, sopping wet, to catch the run. It was a disregard and we immediately returned to the house. I hustled in, still wet, and hopped back in the shower.

When I finished dressing and combing my short, wet hair, I came out and announced over the P.A. the locker room was open. One of the guys who was my favorite volleyball player and was always asking me to make cookies, said he had something to tell me that he was a little embarrassed about. He said he went in the locker room while I was taking a shower. He heard the run go out and knew I was gone, so he just walked right in. He was startled to see me in the shower since he didn't know I had come back. He backed out slowly and quietly but wanted to warn me so in the future I would make a second announcement. He said he couldn't help but notice, I had a cute butt. Laughing embarrassedly, I thanked him for the compliment and the advice. At least his sighting of my butt was at a distance, unlike Naked Butt next to me, close enough that I practically buried my nose in his fanny when I turned my head, my first day in the companies, first thing in the morning.

Mr. Naked Butt and I became good friends, despite the awkward first morning. As we were riding back from a run, he and I were sitting on top of the engine behind the cab riding forward. We normally sat low and rode backward. We were up high, and if we had gone under a bridge, we both would have been chopped in half. We were laughing and yelling at each other over the roar of the engine and the wind.

I threw my arms up in the air and shouted, "I am on top of the world."

As we started backing in, we climbed down and stood, still laughing and talking loudly. The officer, back in the seat from his sick day, hollered for us to shut up so the engineer could concentrate on backing in. We restrained ourselves.

When we all hopped off, the officer asked, "Where is that quiet mouse you used to be? What happened to her? I think I liked you better the other way." Mr. Naked Butt and I just looked at each other and then ran off to set up the volleyball net.

At dinner, the guys wanted to know all about me. How long had I served in the army? Was I once an officer in the National Organization for Women, and a member of the NAACP? Could I fly a plane? Yes, it was all true. My, my, word sure does get around the fire department. They definitely had some good sources for obtaining information because I hadn't told them any of that stuff. Maybe they quizzed my brother or Mandy. Maybe it came from my background check when I applied. Their information was surprisingly accurate.

After a month of bliss at this station, I received moving orders from the chief for my next station. Everybody was coming back from summer vacations and this station would have a full crew the next day. It was time for me to go elsewhere. The men seemed sad to see me leave, especially Naked Butt.

Chapter 10

The mother house was my next stop. This station is huge with our headquarters and all the chiefs running up and down the halls in a separate but connected building. There are six double bays with huge overhead doors on both sides so no one has to back in but sometimes they do anyway. I was detailed to a squad that ran with three people. All of the other squads functioned with two people. Since it was a new environment, I reverted to quiet-as-a-mouse mode. The officer was gone so we had three privates. The two regulars each had about ten years on with plenty of experience.

This was my first encounter with EMS and since it was a busy squad, we barely got our gear on the apparatu before we were dispatched to our first run at a high rise across the street for an unconscious person. We drove there but could have gotten there just as fast if we had walked. We unpacked the tech box with our EMS supplies and hauled out the oxygen equipment, two big boxe which seemed to get heavier as we carried them through the high rise. We rushed to the elevator with a person directing us to the apartment several stories up. The room was located next to the elevator so we arrived in record time. Hurriedly opening the oxygen, the guys motioned me to stop. They disregarded the ambulance over the radio. The police arrived and we left right away.

As we were getting on the elevator, I couldn't keep my mouth shut. I blurted, "But aren't we even going to do anything? Is that it? We aren't going to try to save her?"

Grabbing me by my arm, they pulled me off the elevator and took me back to the room. "Look at her. She's been dead for a while. See the purple blotches on her back, that's pooling of the blood, livor mortis. She is cold and stiff, rigor mortis. There is absolutely nothing we can do for her."

I stared at her looking at undisputable signs of irreversible death, touched her cold limbs, but it was still hard to accept. I felt a lump in my throat growing but I swallowed hard as we quietly left the room heading back to the elevator.

Mistakenly, I thought we would be saving the whole world, working every patient, restarting life. All of our training prepared us for the save. Only an obscure footnote talked about the possibility that we wouldn't be able to resurrect the dead, but I had forgotten that part, recessing it in the back of my mind. These guys had seen it many times already and recognized it immediately. It would be a long time before I would acquire that ability. We had more runs, lots of runs, but mostly falling down drunks that either the ambulance or cops took away.

That night, between runs, I slept fitfully, seeing the dead lady's face; thinking of the guy in the next bed sleeping naked. He asked me if I would mind. Of course, I said no, but it was still weird. I didn't want anybody to have to change because a female was in the house but I worried that I might wake up with some naked guy standing next to my bed. After a few weeks went by, I realized I had nothing to worry about. He was a prankster and a comedian but wasn't going to bother me.

They had a basketball goal at this station and he always picked me for his team. With my outside shooting and inside passing to his tall, lanky ass, we kicked butt. He begged me to play in a league on the department basketball team, but I couldn't because I was busy raising my son. It would have taken too much time away from him even though it was tempting, because I missed playing. I grew up shooting hoops and played college ball but there wasn't time for that now.

This squad had a lot of night runs. At the last station, I watched the squad leave. Here I left on the squad as the others rolled over. This house was a two story building. When a run came in, I got up and slid down the pole which is as fun as it looks. The squad guys excitedly said we were being called to an apartment fire for extra help and I should quickly get on my fire gear.

When we rolled up, there were a lot of apparatuses present with firefighters working feverishly to put out a big, apartment fire. We grabbed a hose line, hooked up to an engine that already had several lines umbilicalled off it and flaked it out to the door. The fire was roaring out a couple of the windows. People stood in the grass, crying and hugging each other. They had all gotten out but it was unclear yet if everybody was out. A search was in progress, and we were assigned a floor below the fire. I had my mask on already, but the other two guys didn't. Abandoning the line, we walked in. It was smoky and dark, but we could see with our flashlights. Banging on doors and busting them open, we found no victims.

By the time we finished our search, the fire was extinguished. The fans were being set up to blow out the thick, lingering, black smoke. We asked the chief for our next assignment. He told us we could go back to the station since the fire was out and there were no injuries to be attended. I asked why we didn't stay to pick up hose. My crew said that squads were the best because they got the fun part of fires, which was putting them out and leaving before the boring work of overhaul and picking up hose began or standing around waiting for the chief to decide if everything was done.

We went back smelling like a fire making the guys jealous they had missed one. As soon as we came into the bedroom, a couple of the guys rolled over and said they could smell us and how was it? It was big talk at the kitchen table in the morning. I'm sure they discussed how the new sub did before I came into the room. I felt pretty good because I had kept up with them and truly helped, making a valuable contribution to the team. I was a little shaky when we pulled up and saw the fire shooting out the windows but quickly got too busy, which didn't give me time to feel fear. The shot of adrenalin my body delivered helped me perform admirably. I figured they would say I did okay for a new sub in a mask.

The next day we got our usual EMS runs. When we got out of the squad for the first one, the acting officer offered to carry the first aid box for me.

The other guy's mouth dropped open and sarcastically he blustered, "You never offer to carry it for me!"

"Oh, come on, we don't get many subs who really care so I thought I'd just help her out."

"Is it the 'caring sub' or the 'her' part that is important? Tell me which it is."

Seeing the trouble developing, I refused to let him carry the box, not wanting to be the cause of a fight, but I felt I had gotten a coveted compliment. I now had an apartment fire under my belt, and someone acknowledging my hard work. I couldn't have been happier. It worked out nicely for me. Starting slowly, I experienced a dumpster fire, progressed to a garage, and became totally indoctrinated with a rip-roaring apartment fire. I couldn't have planned it better.

Chapter 11

It's difficult for me to remember where I spent the next few months except that I was downtown in busy companies where they get a lot of alarm runs, and I had a great time everywhere I was sent. All city-center firehouses are big with lots of guys and numerous activities. I remember one night being in a company where the stuffy chief, who saw me soaking wet my first day, was stationed. A big poker game was in progress. We were in the classic, smoke-filled, parlor room with a round, poker table. Some of the men had cigarettes, some puffed on cigars. The chief had a big stogie hanging out of his mouth that he was rolling around with his tongue. He was losing his shirt but betting heavily anyway. I didn't play much poker, which I had confessed but the guys said the cards played themselves, so if I had a better hand than what I thought, they would tell me after I laid them down. I knew enough that I should stay in with three deuces.

The last round of cards was dealt in a kind of game where you had some cards showing, some cards hidden, and some on the table to play off of. I was transparently excited because I had been dealt another deuce. The chief raised heavily and several matched to stay in, including me. The chief showed his full house and the others threw their cards down in disgust.

The chief started to rake in the money but somebody stopped him and said, "We don't know what Kathy has."

I said, "I just have two pair. Two pair of fucking twos!"

The guys all started laughing and the chief, being flat broke, got up and stomped away, smashing out his cigar as he left. They slapped me on the back and told me it was great to beat him like that because he is tight with his money, getting upset, being a poor sport, if he loses a nickel. I was moved out of his battalion the next day.

I spent some time at a big, north side station where the officer loved loons. They called him Looney Tunes. He had recordings of loons calling to each other and sat outside on the park bench listening to his tape. He told me he imagined being on the water. He would close his eyes as the evening grew dark. Hearing the loon calls would transport him to his boat. Sitting with him on his park bench outside the bay doors, I enjoyed listening to the loons too. The guys probably thought I was a suck-up but he was the first officer I hung out with. He told me a lot of fishing stories and all about his family.

He talked about how he would retire in a few years and move away. He stayed another 15 years. On this job, a lot of the old-timers say they will retire in a year or two but when the time comes, they aren't ready. It is hard to leave. Our lives are tied to the job and it is scary for most of us to think about life without the department.

I followed him into quite a few fires, right on his heels. We were in a rough area of town that had frequent fires. Like the first station I visited, there was a squad at this house. I was on the truck, then the engine, and didn't go on any EMS runs. I wasn't fond of EMS runs after the dead lady.

When a sub arrives at a house, generally the first things they are assigned is to clean the toilets and cook. This house didn't have me do either upon arrival but after seven or eight duty days, it was my turn to cook. Trying to please, I asked if ham and cheese sandwiches would be okay for lunch, directing my question to a couple of the guys standing around.

One of the guys said, "Okay, who put you up to this?" I just looked at him. "I know somebody put you up to this. Who is it?"

I had no idea he was Jewish and didn't eat ham. I swore it was just a coincidence and changed the menu to tuna fish sandwiches and soup. It's funny how one can accidentally cause trouble. The engine officer took the call for me to move. I heard him ask the chief if I could stay longer saying I was a good sub and good subs were hard to find. I hit the road again, carrying with me his words of praise. I was so happy.

Mandy and I shared experiences about different stations. We talked about who was fun, who was grumpy (her shift commander), about the academy, and decided that cutting our hair was no sacrifice to have this career. We would go bald in order to keep our jobs. We were having great adventures, just like my brother said. The days passed quickly and soon it was turning cold.

Mandy went from holey underwear to having her own apartment because of our fire department salary. My son and I went to see her once in a while and she came over to my house sometimes. Getting tired of my straight, short hair, we decided to give me a permanent. We purchased all the stuff, and with Alex jumping off her couch she put the perm in my hair. It turned out pretty good. It was different anyway. My husband hated it but he eventually got used to it.

Christmas was approaching and Mandy was excited because finally she had some money to buy presents for everybody. She was scheduled to work Christmas Eve and I was scheduled to work Christmas Day.

I was at a firehouse on the south side of the city and my chief told me I would probably be there for six weeks, throughout the Christmas season. A lot of the regular guys were off so we had close to as many subs as regulars. A couple of the subs had five and six years on. They were talking about one of the regular guys there who had come on just 18 months before them and acted like he was better than they just because he had a regular station. They were really mad because he had moved up from cleaning the toilets at their expense and all of the subs had to cook before any of the regulars took a turn. They said he was acting mighty high when he was nothing but a near sub. They had to explain to me what a near sub was; so fresh from going regular. The people remaining in the house were younger guys because only the older guys with a lot of time on the job could get the Christmas season off. The younger guys didn't respect their status as senior subs. A lot of the houses they went to treated them like regulars because they had so much time on the job. We subs hung out together, scorning the regulars, since the senior subs were fed up with the near subs.

One evening the guy with five years on said to the six year sub, "Come on, let's go find us a woman."

The six year sub said, "We got Kathy right here!"

"Not that kind of woman. I mean a real woman we can bring back here and take upstairs."

There was a bar down the street from the station that the guys walked to with a radio in their hands. Since I would have hampered their chances of picking up a woman, I wasn't invited. Plus, I was too new and too chicken to tag along so I stayed back and played poker. After all, I was still probationary and could be easily fired, according to my instructors. There were vague rumors about maybe that happened to someone in IFD history but nobody could really remember when or where or why. Still, I didn't want to take chances. I had already heard about one of my classmates missing a run in the middle of the night. According to the rumors, he slept on the apparatus now so he wouldn't blow any more runs. I didn't want to have a beer or two, especially since I can't hold my liquor, and have the whole fire department hear talk that I was in a stupor and missed a run.

Some women workers from the ice cream store came over after it closed. They and a couple of the guys disappeared upstairs. After about an hour, they all came back down and watched us play poker, laughing at the jokes being told by the guys around the table. Getting late, the women went home and a lot of the guys went to bed. My sub friends came back empty handed and were distraught to discover they missed the "fruit broads". Now they had to explain fruit broads to me.

"The kind of girls who hang around the firehouses and are very friendly. You know. They call them fruit broads because the picking is easy."

"Ohhhhhh!"

That was more than I wanted to know because they seemed like nice women. They were happy and friendly and the one woman seemed to genuinely like the one guy. The only problem was that everybody, including the women, was married. Some of the wives of the firefighters fussed when women were hired on the job. They had more to fear than us. Perhaps they had no idea this other stuff was going on, or maybe they were in denial.

I still exercised every work day and made my announcement on the PA when I was ready to take a shower. This house had a lot of guys who ran and lifted weights. Since the time was short to work out, shower, and dress, I would wrap a big towel around me after I cleaned up, get out of the locker room, and dress in the public restroom. I passed the guys going in as I was going out, bundled up in my towel.

One day the six year sub said he was going to be me and announced the shower was open. We were hanging out on the stairway as he exited the shower with a towel wrapped around him. It was a micro towel, barely covering his chest and belly button, letting everything else hang out. As he swung his hips and limp-wristed around, we were all rolling on the ground laughing as he imitated my walk with his big, hairy, butt showing. He almost did not need a towel because everything was covered with furry hair, resembling some kind of missing link. I gawked because I had never seen anyone so hairy.

On my next day off, Mandy invited me to her place so we could exchange gifts. Alex had to sit on the couch because the people downstairs complained the last time we were over when he jumped off her couch, thudding on the floor. She gave Alex a little toy. For me, she flourished a hooded, white sweatshirt with blue trim. Across the back in matching blue letters was: FIREFIGHTERETTE GILLETTE. I put it on right away and wore it until it was holey, just like her underwear. It meant a lot to me and I have never forgotten that thoughtful present.

My family opened Christmas presents on Christmas Eve since Steve and I both had to work Christmas Day.

We reported extra early Christmas morning so the C shift guys could go home to their families. I left Alex tucked snugly in bed with his new toys scattered all around him, some cradled in his arms.

It was quiet at the firehouse because the guys were leaving immediately upon relief. We were sitting around the kitchen table, our shift all there, when an apartment fire with entrapment was reported. It was a very, cold day and I was glad I had on my new, long underwear. The guys were teasing me because they were white with little, pink and green flowers. Most of them wore dark blue, long underwear under their short sleeve, blue, uniform shirts. They were a Christmas gift from my mom who likes cute, feminine stuff. Mine were brand new out of the package and still very white because I hadn't washed them yet in cold water with everything thrown together. Not having time to sort resulted in most of my whites turning a tattletale gray.

We all rushed to the apparatus as another radio announcement stated control was receiving multiple calls for help. When multiple calls are reported, you can rest assured it is a real fire and not somebody imagining the smoke from a chimney is a house on fire. It wasn't far from the station, just around the corner. I was accomplished getting into my gear now and was fully dressed before we arrived. Seeing the smoke as we approached, my body pumped up with adrenaline as the other sub hopped off to catch the plug. We pulled up in front of the building with heavy fire showing on the top floor. People standing on the street frantically motioned and yelled that there were old people still in the building. As the acting officer pulled off the hand line, I helped. From the sidewalk, he opened the line, spraying water at the window. The six year sub came up from catching the plug and grabbed me.

He said, "That is no way to fight this fire. Come with me. We'll knock it down."

We pulled another line and dragged it into the building to the top floor where the fire was. Huddled in the corner of the stairwell, barely visible in the smoke, was an old couple. We opened the door to the apartment and the flames rolled toward us. He opened the line, hitting them and started knocking down the fire. The smoke billowed out making it difficult to see. He inched his way in, keeping lower than the smoke because he didn't have a mask on. I could feel the heat through my gear. As he advanced, he shielded his face from the heat with his arm and started coughing his lungs up. I wondered how he could stand breathing that crap because I could barely stand it during the clean up when most of the smoke was gone. My eyes water and I get a snotty nose.

He yelled at me to get the old couple out of the stairwell who were crouched together in the corner, barely able to breathe, crying and scared out of their minds. They would soon be overcome by the smoke and we would be hauling them out on stretchers. He said he could get the rest of the fire out by himself.

Grabbing the two people by the wrists, I pulled, pushed, and urged them to scoot along the floor to the stairwell. We descended a few steps and the smoke thinned out. They were able to stand now with help. More firefighters were coming up the steps to assist. I handed them over to the arriving firefighters and went back to the fire room where my sub buddy was stranded. Crawling in, following the line, I found him lying on the floor putting out the last of the fire. It was a small apartment so it didn't take much longer. We shut down the hose line and crawled back out as the truck crew ventilated the place to clear out the black smoke.

As we went down the steps, I took off my mask which was low on air. Reaching the last step, I turned around to wait for the other sub. Descending out of the haze, I saw his face. Hanging out of his nose was a six inch long string of snot. His face was sooty and smeary where the tears from his eyes ran down his cheeks. I told him how disgusting he looked. He shook his head, swinging the snot from side to side. I gagged. He laughed. He showed his trick to everyone else who mistakenly looked. Then he wiped his face and lit a cigarette.

The five year sub, who had been very nervous about driving and had started the engine in the morning about five times to practice pumping, did an excellent job of getting us water. We slapped him on the back, congratulating him. We gossiped about the regular guy in the seat who wouldn't go in to fight the fire. They were incredulous that he was a chief's son, nothing like his fearless father. They argued, placing bets on how much longer he would be on the job. They had him pegged. Afraid to go in, he quit a few years later.

Cleaning up after a fire takes a lot longer than putting it out. If there is a question of arson, we can't clean up much because we destroy evidence. Arson has to be called. We wait for their arrival and they investigate a little or a lot. Then we are set loose to throw out everything that burned, making sure there won't be a rekindle. That usually meant we had to tear out some walls and pull down some ceilings. If there is attic access, we climb in there via a little, folding ladder made to fit in the narrow openings. I get sent up into a lot of attics since I am smaller than most of the guys and can easily fit in the openings. It is always very hot crawling around up there and I get covered in insulation that sticks to my wet gear. If there is any fire, they pass a line up and I give it a squirt or two to douse the flames. Then I come down looking like a ball of fluff. Today, tearing out walls is generally unnecessary because we have thermal imaging cameras or TICs that picture hidden heat. We tear out a wall only if a hot spot is detected by the camera.

Since there was no attic access and the origin of the fire was determined to have started because a cat knocked over a candle that caught the curtains on fire and spread quickly from there, we were loosed like the hounds to destruct, tearing down ceilings and chopping holes in walls. My sub friends were outside smoking but I was still excited about every aspect of the job and helped chop and tear. Plus, it was a lot warmer in there than out on the street where icicles were forming on the butts of our hoses and the coats of our jackets. The buckets were brought out with warm water from the pumps to dip the hose butts, melting the ice so we could screw them back together after we drained them.

We finished the clean up and the pick up heading back to our cozy firehouse. We were enjoying lunch, the six year sub remarking how green stuff, like salad, is good for you because it makes your poop slick and prevents hemorrhoids, when the Centrex rang. We were told the mayor was going to come by and congratulate us for a job well done and saving lives. We had about an hour before his arrival. Cutting lunch short, we washed the dishes, rewashed the apparatuses and lined up for his arrival. The acting officer, who wouldn't go in, said I should take off my white, flowery, long johns before the mayor arrived because they weren't appropriate. Politely refusing, I said I wouldn't freeze my ass off for anyone including the mayor and he would just have to understand.

The mayor arrived with a big smile, shaking everyone's hands as we stood in formation. He held my hand extra long saying he heard about my save and what a fine job I was doing. He added as he winked, he really liked the flowered, long johns, and then moved on for the next handshake. After he left, we laughed about his long john joke and decided he was a pretty, nice mayor to come out on Christmas Day when he didn't have to. It wasn't a big media event either, just a grateful mayor and his aides.

We had a really, fine dinner together with some family members stopping by to wish us a Merry Christmas. Every year for Thanksgiving, Christmas and Easter, some local corporations donate money to the firefighters for their holiday meals. We use the money to make big feasts for the special occasions. Usually there is so much left over, the shift reporting the next day doesn't need to cook. This station had a permanent cook so I didn't have to worry about a cooking rotation. He was on the busy squad and I wondered how he had time to shop, cook, and take runs. He was dedicated though. Most of the time, he would stop at the store in the morning on his way to work and still be one of the first to arrive. It is a tradition on Sunday mornings to have breakfast and he cooked everyone's favorites. Still, the guys teased him mercilessly, which I never understood. I always complimented him on his cooking but some of the guys complained. I thought he ought to tell them that they could cook their own damn meals if they didn't like it. After dinner and all the family members left, we had a quiet evening. The squad had a few runs for turkey indigestion but not much else.

Cooking in the big houses is quite a challenge. Just lugging out the enormous pots suffices for a weight lifting session. All of the shopping has to be done in the industrial section of the store. The stress of getting the meals cooked on time and within budget is tremendous. It is akin to big, holiday dinners when you start out thinking you'll impress the out of town guests but in the end, you are glad to just get the turkey on the table after you have thrown out the burnt rolls.

There had better be leftovers too because if there isn't enough food, called laying out short, retribution is swift and certain. I heard a cook could be painted for such a grievous offense. Painting is the act of stripping a guy naked, then spray painting his testicles while he is held down. I never saw it happen but I heard about it many times. A time or two, the guys threatened to paint me because I could be obnoxious with my excess energy, jumping on beds as they tried to sleep, wanting them to get up and keep me company. Diffusing the situation, I told them they would have to get tweezers. That made them stop and think, then laugh.

The shift that works Christmas also works New Year's Eve. We were stuck at the firehouse and would miss the revelry but firefighters seem to make the best of every situation. Mandy came by with my favorite, a bottle of peach schnapps. The girls from the ice cream store stopped in, joining the party. Dancing and celebrating the

New Year at midnight, we toasted with champagne the guys pulled magically out of their lockers.

Shortly after the start of the new year, the regulars returned. Feeling quite at home now, I was sad to leave, but all good things for new (and old) subs must end.

Chapter 12

My next big adventure happened when I was sent to the north side of town to a company with an engine and a truck. The lieutenant on the engine was recently promoted. It was to him and his apparatus that I reported. He had six or maybe seven years on, which is a very young lieutenant on IFD. In the afternoon, an apartment fire in a run down government complex was reported. Box alarms or extra companies are automatically dispatched on reported apartment fires because of the possibility of disaster. Most of the reports are false or minor incidents like food on the stove or someone barbecuing on a balcony. Since we saw black smoke in the sky while on the way there, we knew it was more than that. Suiting up, I got my big, adrenaline rush and was ready to go. The engineer stopped at a hydrant. Being a young sub who needed experience, I stayed on the engine until we were in front of the fire.

The officer got off and put on the hose clamp as the engineer was engaging the pump preparing to throw water. Jumping off, I started pulling a line that was preconnected to an outlet. The officer, now ready to assist, helped get the line over our shoulders and purposely dropped portions of it as we went. The truck crew pulled in front of the engine, parking head-on to the apartment, forming a t-shape with the engine. If they needed to raise the aerial ladder out of its bed, it would be in the most stable position, in line with the chassis. If we moved to a defensive mode, they were in a perfect position to spray the apartments with a powerful stream of water delivering hundreds of gallons of water per minute. Fire can easily spread, and get out of control in buildings that share a common attic. They were set if such a grim alternative took place.

The new lieutenant and I kneeled at the entryway, donning our face pieces and turning on our tanks. The engineer, upon the lieutenant's signal, opened the valve giving us water. The line stiffened with the 120 pounds of pressure. The ascending steps were on fire. Hearing the familiar hiss of escaping air as the lieutenant opened the line, the water shot out and we doused the carpeting on each step as we walked up, putting out the fire. When we came to the doorway of the apartment on the right side, the heat became tremendous. The door was already opened. The lieutenant dropped to his knees and started crawling in. The hellish heat, now that I was no longer shielded by the big lieutenant, drove me to my knees and I landed on top of his boots. With fire all around us, the lieutenant was whipping the hose around in order to knock down the most fire possible. We were making good progress putting out the leaping flames. Everywhere the officer sprayed, the fire went out. We had doused about a third of the fire when the officer turned and said he couldn't get air. There was something wrong with his mask. He shouted through his respirator that I should take over the line. He told me to continue putting the water on the fire, and he would either be back, or send someone in to help.

As he left, I moved to the nozzle, opened it and started whipping it like he had, spraying to the left, then right, and up and down in front of me. It was considerably hotter now because I had no one in front of me to shield the heat. Moving forward as the flames went out; I was approximately half way into the room. Suddenly, the water shut off. What in the world did I do wrong? I pulled the handle toward me to shut the water off and then pushed it away to turn it back on. When nothing happened, I yanked the handle back and forth. Shaking the nozzle and banging it on the ground did not help. The line became limp. I looked into the nozzle which was stupid, sort of like looking down the barrel of a gun. If it had finally started spraying, it would have knocked me over backward, but no such luck. Had we discussed this type of situation in the academy? Searching my mind, I tried to remember. What could I do to get the water flowing again before somebody discovered my goof? My only solution was to yank the handle some more. Hearing a lot of yelling behind me, I was snapped out of my thought process. When I looked around, almost all of the flames we had put out were leaping back up around me. I was becoming surrounded by flames!

The guys in the doorway were motioning to me and yelling, "Get out! Get out! Get out!"

Throwing down the line, abandoning it, I crawled as fast as I could for the door. Standing was not an option because the heat was unbearable and it would have melted my head. There can be a thousand degree difference between the floor and the ceiling in a fire. Hot embers were burning through the pull up part of my boots. As I reached the door, the room flashed over, erupting in flames. The guys at the door ducked down as fire shot over their heads, flew across the hall and flamed like a blow torch against the next door neighbor's apartment door. Groveling on the ground, they reached for me and grabbed me by my tool belt, flinging me out of the room so violently that I almost went down the stairs headfirst. Recovering ourselves, we crawled down the steps, listening to the flames roar across the hall and the glass in the windows erupt.

Exiting the building, we saw it was bedlam outside. My water supply was lost because the old pumper had jumped out of pump gear, lurched forward, and crashed into the truck. A firefighter was counting his lucky stars, because he had just passed between them, barely missing being pinned. It surely would have cut him in half, because the truck step was deeply dented where the bumper of the engine had dug into it.

The chief was screaming on the radio for an alternate water supply because the engine had torn the hydrant from the ground, and it was now unusable, spraying water into the air. The apartment next door was starting to light up as the flames burned through the door and spread. Pulling down my boots, my knees were beginning to blister. My face, neck, and ears felt burnt. The chief saw the blisters on my knees and ordered me to go to the hospital. Not wanting to go, I argued with him but he wouldn't listen. He said I had to get checked out. The medics loaded me into an ambulance and took off for the hospital. Enroute, the medics poured water from a saline bag over my knees to cool them. Their kind efforts helped.

Dumped at the hospital, I waited in a small room, worrying about the guys back at the fire. I was seen by a doctor after approximately an hour. The doc said I had mostly first-degree burns, something akin to a sunburn. He was concerned that the blisters on my knees might leave some scars. He told me if it would bother me, I could request plastic surgery to remove the scars. It seemed ridiculous to worry about a few scars on my knees. If they were on my face, that would be a different story. Scars on my knees would be something to brag about, an unusual conversation piece.

The doc gave me some burn ointment and a prescription for an antibiotic cream that I had filled before I left. The safety officer picked me up at the hospital and delivered me back to the station. Dirty from the fire, I took a cool shower, because my burns were sensitive to heat. The cool water felt good on my knees. Toweling off, I put the ointment on my burns and wrapped them in dry, sterile bandages.

The Centrex rang for me. It was arson. The investigator, a firefighter who had been a guest lecturer for my class at the academy, wanted to ask me some questions about the fire. He needed to know the sequence of events, the color of the fire, what exactly was burning and other miscellaneous questions that I can't remember. Being inexperienced, most of the questions were beyond me. The majority of the fire environment, I couldn't remember much about it. I doubt if I was very helpful. The lieutenant, close by, heard the answers I gave. He was upset and told me he didn't want me to repeat to anyone else that I was in the room by myself because he could get into trouble for leaving me since I was still probationary. Not wanting to get anyone into trouble, I didn't repeat that part of the story anymore.

In order to miss the lecture from my mother about finding another job because this one would be too dangerous in her opinion, I didn't talk about the alarming incident at home. It would have scared the pants off Mom and like a rabid dog, once she sinks her teeth into something, she doesn't let go. She would have the newspaper out, circling all the desk jobs that were safe. Sticking them under my nose every time I visited, I would have no choice but to look.

To tell you the truth, even though it is a dangerous job, I felt safe. Surrounded by great firefighters who were watching out for me, I knew I would be taken care of. Even though it was a close call, my buddies were there watching my back. I am grateful to the firefighters who were cognizant of the imminent signs of flashover, and knew I needed to bail out. I had been too busy trying to get the water back and failed to pay attention to my surroundings. The firefighters who pulled me out by my tool belt saved my life.

At dinner, I heard about the rest of the fire. It was discussed and dissected. An alternate water supply was located. Since the hydrants were far away, the water was relay pumped by two engines. The fire was confined to the two apartments. No one was at home, so there were no injured civilians or crispy critters. (Crispy critters are bodies burned beyond recognition. It may sound harsh, but it is a commonly used euphemism in the fire service, part of our strange humor. In fire school we were given a list of slang terms used on IFD and crispy critter is on it.)

The question was raised as to why the engine jumped out of pump gear. The engineer, defending himself and throwing his hands up as if he were being robbed, said he put the chained hook on the pump engagement, pull-out handle just as he always did. This particular model was infamous for vibrating out of pump gear. The special hook was devised by the mechanics and placed over the extension rod by the engineers to ensure the vibrations wouldn't rattle the rod out of pump gear. It sounded to me like it was not a perfect fix. The mechanics would have to go back to the drawing board.

Chapter 13

Substituting at all of the companies, I gained valuable firefighting and EMS experience. I was also treated to my share of good-natured hazing. Being taught early on to make my bed in the mornings, it was frequently targeted for pranks. Short sheeting the sub's bed was a favorite past time for the regulars. The first time it happened, and I couldn't get my feet but halfway down the bed before they got caught in the sheets, I was puzzled. Slow to figure it out, I attempted re-entry without success. Since I went to bed late, it was difficult in the dark to take the sheets off and remake the bed. Under my breath, I cursed the trickster.

Several times my bed was floured. Because my feet were always cold, I wore socks to bed and didn't feel the flour at my feet or notice the white footprints I left as I got up to use the bathroom.

One night my face was tickling. When I scratched, I jumped out of bed, brushing off a spider. Standing in the dark shivering, I heard the muffled chuckling. Turning on the lights, I saw the fake spider dangling from a string attached to the ceiling. Another night, those same guys threw a fake rat on me, but I was wise to their games, spoiling their joke.

Another popular trick is to put a rubber band around the kitchen faucet sprayer. The subs, first to jump up to wash dishes and turn on the faucet, would inevitably take a shower a couple of times until they learned to check the sprayer for rubber bands. Out of habit, I check the sprayer, even now.

I began revisiting some houses where I had happy reunions with friends I made on prior visits. I was sent back to the company where I spent Christmas. The cook continued to be razzed. I met the regular guys who had been off for vacation. The weather was getting warmer and I was approaching the end of my probationary year and received excellent reports from the officers for whom I worked. The men at this south side company thought I was a good subbie and everything was going well.

My husband, still in school, needed to be early for ROTC one morning. He called the night before asking if he could bring our baby by the station in the morning on his way to class. I got up and checked if my relief man had arrived. He relieved early, generally around 0630. It was 0700 and he was in the kitchen drinking coffee at the crowded table.

I met my husband in the parking lot, taking Alex with me into the firehouse. He was standing on the bench as I introduced him to everyone when the captain from the B shift butted in asking me if I was relieved. I said yes and he asked who relieved me. I said the same person who had been relieving me for the past four weeks. He asked me if my gear had been taken off the engine. I said I did not know. He motioned me to follow him to the apparatus. I followed and he pointed out that my gear was still hanging on the engine.

"Your regular relief man was moved to the truck and your gear is still there so you have not been relieved. You are derelict in your duty by not being able to catch a run if one should come in."

I stood dumbfounded, my heart thumping.

By now, several of the guys from my shift were coming out to stick up for me. One of them said he was relieved and would relieve me. He got his gear, put it in my spot and took mine to the rack. The cook lieutenant stepped between me and the captain and told me to go home, that he would talk to the captain. I left, surprised that someone would show that kind of animosity toward me. I think the guys on my shift were surprised too. I got out of there and did not think much of it because the lieutenant told me not to worry.

When I reported for duty the next A shift day, I was pulled into the office by the lieutenant. He was headed downtown because he had to give an official report of the events that occurred when I thought I had been relieved. He said the son-of-a-bitch captain hadn't wasted any time going straight downtown in the afternoon, disciplinary report in hand, charging me with dereliction of duty and conduct unbecoming a firefighter. Either count could get me fired. The lieutenant said the report was exaggerated and he would straighten it out. My stomach was churning much worse than it had on any fire.

After a long, worrisome morning, the Centrex rang for me. The lieutenant told me the charges were dropped and everything was okay. The A shift chief came on the line next, saying the captain had it in for me so it would be best to move me away from him. I packed up and was at another station before lunch. The men seemed sad to see me go. I was disappointed too because I was scheduled to be there another two weeks. I genuinely liked the guys there and was sad to leave with a cloud over my head. The lieutenant said he thought the captain was jealous because everyone talked so highly of me.

I was sent to an outlying company that was fairly quiet. The area consisted of small, single family, homes with a few apartment complexes. Having spent the majority of my time in fast companies, I was going a little stir crazy. Reading the bulletin board for something to do, I discovered an official posting for an upcoming engineer school. It said anyone non-probationary could attend and applications were being accepted. Since my probationary year was about to end, I considered applying.

The next day my chief called to ask how things were going. I told him I missed the action in the busy companies. He said all I was missing were a lot of false alarm runs. In this company, he said, they may not have tons of runs but what they get are usually real and important, and I should just be patient. When asked about engineer school, he encouraged me to attend because he did not have many sub engineers and could use some. He also told me that an engineer could go regular as soon as a position could be won through seniority. If I went to school and was certified, I could go regular ahead of my class. I would also get paid extra any time I drove an apparatus. He said he would tell the lieutenant to let me practice-drive on the way back from runs because a requirement for entrance to the school is logging a specific number of practice hours certified by an officer. Even though I would have to go on my own time, I was convinced attending would be a good thing. I asked my mom if she could watch Alex for me during class. She heartily agreed since it would mean more money for me in the near future.

Before I was off probation, I started driving the engine and truck back from runs. The hours were accumulating slowly though since this was not a busy company. I convinced the officers to go for drives around the neighborhood so I could rack up more time. They fudged the hours just a little because they did not particularly like missing their afternoon naps to let me drive.

The engineer on the truck gave me pointers on driving. These behemoths were daunting pieces of equipment. I drove some smaller trucks for a summer job after I got out of college but nothing like this and not in an emergency situation. I was nervous but he reassured me. All the talk got him fired-up about driving and the next run we responded to, he drove like a bat out of hell. His truck was older and did not have a cab area where you could sit backwards on the way to a run. Backsteppers were required to stand on a step by the ladder just behind the cab. To hang on, we wrapped an arm through the rung of the ladder and grabbed a wrist with our other hand, holding it against our body.

On most apparatuses, we don our gear on the way to the run. On this aerial, we got dressed before take-off so our gear would not fly off the truck onto the street as we hastened down the road. The sirens were ear-piercing loud. The wind blew the sound right past our ears with nothing to block it. This company had a large district. This particular location seemed a distant ride with the other sub and me hanging tight to the ladder. I could see him standing on the other side. We tightly buckled our chinstraps or our helmets would have flown off.

As the engineer entered the gravel lot of the business, he veered right and hit a chuck hole. My feet flew up and over to the side, smacking the ladder then thumping back onto the step. The poor sub on the other side was hanging straight out with nothing to stop the swing of his legs. He looked like a pennant in a stiff wind. When we started slowing, he was able to maneuver his feet back on the step. We bounced in the air a few more times as we hit more chuck holes.

Jumping off, we investigated and disregarded the other companies because it was only popcorn in the microwave. The officer chewed the engineer's ass for practically throwing the subs off on nothing but a false alarm. I drove back listening to the officer highlight the duty of an engineer to transport everyone to the scene safely. Speed was not as important as safety.

The chief was right. This station, even though it did not get a lot of runs, had several fires. We put them out before other companies arrived. A government subsidized, two story, brick, apartment complex was in this company's district. Firefighters have to break down doors frequently to get to patients, or to enter places that are on fire, or to investigate gas odors. Sometimes neighbors will have keys. Sometimes the doors are unlocked. I had already broken down a few doors. I have a padded hip that makes a great battering ram. I slam it against the door a few times and pop, the door is open.

We were dispatched to this complex for an apartment fire. When we arrived we could see flames through the picture window. We dragged the line to the locked door. I gave it a couple of whacks with my butt to no avail. I knew it was going to be a tough one to open but the lieutenant wanted to give it a try even after my warning. He kicked the door with the bottom of his boot in front kick fashion. The door did not budge and he jammed his femur just about out the back of his leg. He had to sit down while the truck crew went to get a battering ram.

The flames were multiplying fast. Leaving the lieutenant, I picked up the deserted hose stiff with water lying on the ground and swung it hard, smashing out the picture window. The flames leaped up with the fresh oxygen supply but I opened the hose line and extinguished them. The guys returned with the battering ram, knocked down the door and we entered.

Upon entering, we searched the residence and found the fire was contained to the front room. Someone had thrown a wad of clothes in the corner and ignited them with gasoline. We set up the fans, cleared the smoke, picked up the hose, and drained it. Arson arrived. After being released, the lieutenant hobbled to the engine. He limped for the rest of the month while I was there and for a while after that, from what I heard. When I returned six or seven months later, he was not limping.

One of the guys on the B shift at this station became one of my best friends. He is retired now but comes to my firehouse for frequent visits. He was always pinning me down to talk. I learned all about the woman he married, who was also his high school sweetheart, his children, his day-off job, which was repairing grocery carts, the volunteer work he did, and his farm. He owned ponies which he took around to different sites and charged people to ride them. He donated the pony-money to charity. He needed people to lead the ponies so I helped whenever I could. When we didn't have kids waiting in line, my son got to ride the ponies too and he loved it. My friend drove the truck, offering me lots of inside information about being a good engineer.

He cornered me so frequently, the guys would yell at him, "Let her up! She's got a kid to go home to!" (After the incident at the south side station, my husband never dropped the baby off at the firehouse again. If he needed to go somewhere early, which occurred frequently, he would take Alex to my mom's.)

The next station I reported to was on the west side. It was situated in a lonely, industrial area. I asked for directions to get there. I grew up on the south side and didn't know the west side.

When a sub asks, the regulars say, "You can't get there from here."

Laughing politely and waiting patiently for an answer, they gave me directions which I wrote down on the back of a dispatch sheet.

The air was dusty from dump trucks and big rigs driving out of factory yards, packing tan dirt on the roads. The nearest grocery store was miles away.

The lieutenant was a friendly person who laughed a lot. He let me drive everywhere except to runs. I motored all around the district in the engine including the distant store, stacking up the hours.

One of the men, who helped train our class in jump rope physical fitness and sold his special jump ropes to my class, was there. He trained every morning, whipping his ropes around on a wooden shelf he built to avoid jumping on concrete floors. He played fast music on an old record player, jumping to the rhythm. I retrieved my rope from the bottom of my duffel bag and had fun training with him.

The lieutenant, like me, did not take the typical, afternoon, firefighter nap. Instead, we put on records, crooning along with the West Side Story album. Although the movie was a little before my time and I had never seen it, I was crazy about the music. "Somewhere" was a favorite of ours, becoming a hit again years later when Barbara Streisand sang it on her Broadway album. There was no drinking or womanizing here because the officers were clean cut.

One of the guys liked to bet on the ponies at the race track so we heard all about his adventures in Louisville and Cincinnati. I think he worked to support his gambling habit. He smoked like a chimney in the watch room as he listened to the horse races so we did not go in that room much. Tobacco use is banned in the firehouses now, although a lot of the guys still chew, spitting in cups then dumping them in the trash can. Some of them spit in the kitchen sink!

I was at this station when I completed my probationary year. I felt the relief that my brother felt a few years earlier when he came in the door at Mom's house, announcing he would no longer have to worry about getting fired. My class talked about getting together to have a party but we never did. Mandy and I celebrated by going to a union meeting, drinking peach schnapps, and afterward, meeting at a local bar that played 50s and 60s music. When they put on "Mony, Mony", being inebriated, I got on the dance floor gyrating, doing cartwheels, round offs, backbends, and sky high kicks. Many of our union members danced with us, and even as crazy as Mandy and I acted, we did not sit much all night. Helping myself to one of Mandy's cigarettes, I started puffing on it because I do stupid stuff when I'm drunk. It was getting late and Mandy disappeared with somebody so I went home.

When I climbed into bed, my husband sat up and growled, "You've been smoking! The rest of the world is trying to quit and you are trying to start!"

"I just had a couple of cigarettes while I was drinking. I won't start again!" I promised him and myself.

I had a bit of a hangover when I came to work the next day. I'm pretty sure I only had a few beers. My brother used to tease me that I could get drunk just passing an empty beer bottle. However many beers I had was too much and I was paying for it. Oh, I forgot about the swig or two of peach schnapps!

After a sluggish, jump rope, workout, I felt too tired to sing with the lieutenant, choosing instead to recline and watch TV.

Dispatched on a false alarm, we were standing against a wall waiting for the completion of the investigation. I started feeling a little queasy and slid down the wall sitting on my haunches. Finally released, the guys helped me up and commented that I looked a little green around the gills. Being gluttons for punishment, Mandy and I decided to go to union meetings more often. They are held once a month, on two nights, Tuesdays and Wednesdays, so all firefighters have an opportunity to attend. Mandy and I went on the B shift night, since I was A shift and she was C shift.

Engineer school started. Thanks to the lieutenant on the west side who let me drive everywhere, I had plenty of driving time to qualify. To give firefighters from all three shifts a chance to attend engineer certification schooling, three days of classes were held. A firefighter could attend class on either of her alternate shifts. If someone taking the class had a busy shift where she ran all night, she could go home instead of falling asleep in class which was a no-no, and attend refreshed the next day.

I primarily attended B shift days. Mary, my academy classmate also attended. Classroom training and driving instruction was about 50-50. We had to pass the classroom tests in order to move on to the driving. We learned about water, its properties and weight, the dynamics of pumping, and hydraulics. It was technical, interesting and informative. There was a special session for high rise pumping and operations because gravity affects water flow. Since we were attending off duty and not getting paid, we wore civilian clothes. Some of the people attending had 10, 15, even 20 plus years on the job. The same instructors who taught me in the academy were teaching this course. They acted more like teachers than hard core drill sergeants.

Finishing the classroom portion, the fun began. We drove every type of apparatus the fire department owned, in particular the aerials and the engines. Some were manual transmissions. My first car was an old, red Rambler, with a stick on the column, which my father gave me. I drove a few deuce and a halves in the army. With the hours I drove in the companies, I felt confident. Some of the would-be-engineers had no experience driving standard transmissions and were having difficulty learning to shift. We met at Fort Harrison, an army base that had recently been downsized. There were long, deserted, stretches of road where we could train and not take out any citizens.

We reported one morning to excited instructors because the department's new turbo-charged engines had arrived from the factory. They smelled new. They were shiny and beautiful without scratches from trees in alleys or dents from accidents. The silver, diamond plate sparkled without the inevitable, calcium deposits from the hard, Indiana water. All the oil-filled gauges had glass coverings and registered properly. Our department never had such powerful engines and we would get the pleasure of finishing school in them. Jumping in, I got to drive one of them first. How lucky can one person get?

As we headed down the long, straight road, the instructor said, "Put the pedal to the metal and see what this baby can do!"

Gladly, I stomped on it. We were whooping and hollering when I saw red lights in my rear view mirrors. We were close to the rest of the class and I had already let off the gas, hearing the powerful chug of the jake brake as it slowed us. After consulting the instructor about what I should do, I pulled over and watched two military police officers (MPs) get out of their jeep to approach the engine.

Coming from the opposite way, the academy chief hurriedly walked towards us. When the MPs reached my window, I told them I was a student and they should talk to the officer on the other side. Not budging from their spots, they demanded my license which I did not have on me. The officer got out of the apparatus, joining the MPs and the chief. They were all arguing under my window. The officer took the blame, telling the chief he told me to open her up. The MPs still wanted to give me a ticket. The chief dropped a couple of colonel's names and after much conferring it was agreed that we would keep the apparatuses at speeds less than 40 mph while on the grounds. They warned us, the next time tickets would be issued.

When we parked the engine, everybody in class started punching me in the arm and laughing, joking that I almost made history by being the first person to get a ticket for speeding in a fire engine.

The next class day I won an award. We were practicing maneuvering around some of the cul-de-sacs to get a feel for cornering. Most of the students stopped and backed up to go around the bend. I decided I would try it in one maneuver. By taking the outside corner as I approached the bend and turning hard early on, I was able to keep the wheels on the road and still make the turn. When I got back to the start, some of my classmates who witnessed the fancy navigation, clapped as they laughed. One of them went to the back and pulled a leafy tree limb out of the ladder at the rear of the engine. I clipped the tree limb going around the corner and did not see it flapping in the breeze in my mirrors. They presented it to me in a mock, coronation ceremony as the queen of corners. The instructors inspected the engine for dents and scratches.

The final test day came with cones to: maneuver around, back into, stop on a dime in front of, and drive alongside. We were awarded points at each test station based on distance from the cones and the speed with which the exercise was completed. I got maximum points at most stations but I goofed badly on one. If you touched a cone you received a zero, making it next to impossible to pass the course. At the stop-on-a-dime-in-front-of-the-cones station, I was too far away to earn many points but it was just a practice time for me. I felt like I had it down pat so I called for a point session. Taking off, rolling, then hitting the brakes, I stopped touching all three cones.

The instructor who is one of the nicest men in the world said, “That was your second practice try, right? We will make this next one for your points.”

I said, “Yes sir. That was just a practice. Here comes the real thing.”

I stopped inches away this time and got five points when I should have received a zero. All I needed was one more point to pass the driving course and the class.

At the next testing station, I backed in fast keeping my distance from the cones. I got two points and danced around with joy, knowing I had just earned my certification as an engineer. The lieutenant, who had always flapped his collar in school because he had no bars but had finally been promoted, was flustered that I backed so fast and hopped out dancing. He measured a second time to make sure I passed, shrugging his shoulders and banging his head when he realized it was true. He asked the chief if he was going to unleash me on the unsuspecting public and the companies.

The chief said, “Well, she passed, didn’t she?”

Mary passed too. We were hugging each other after she finished. He just walked away. Unfortunately, several firefighters did not pass. Smooth operation of the clutch on the tricky backing exercise eluded them. They were not as lucky as me to get the nice instructor when they made their mistake. He winked at me as he left. Since he and I never had an opportunity to discuss the second chance he gave me, I can only guess why he did that huge favor for me. I’d like to think it was because he sympathized with me when my classmate resigned and thought it unfair that I alone had to raise the 35-foot ladder in the academy. He paid me back for that injustice. We all went out and celebrated by having a few beers.

When I got home, the house was dark so I went to Mom’s and picked up Alex. She said my husband called to say he would be late again. I was very animated from the beer and happy to have my certification. Alex was so cute toddling around. We had dinner with my parents. When I got home, my husband was there eating leftovers. He said he had a special ROTC project due at school and had to stay late. I told him all about my test and certification. We had a good laugh about “Lieutenant No Bars”.

Chapter 14

When my certification was officially registered, I was frequently assigned to drive in the companies. The first place I drove was where I beat the chief at cards. The men appeared unhappy as I arrived to drive the engine. None of them had ever been chauffeured by a woman. A back-stepper was normally moved to the front and paid to drive if his engineer was gone but I got the money instead.

One of the first runs where I chauffeured was on a high rise alarm to a bank building. I asked the officer which way to go. He stared straight ahead just like my instructor at the academy when I dropped the pry bars. Pulling onto the approach, I looked at the skyline for the tallest building and headed toward it. I got us there without his help. The next run, I asked the truck engineer for directions and he told me. After a couple of runs, the officer relented and helped me. I was surprised because in the past he was very nice to me. They laughed hard when I beat the chief but they were not laughing now.

For the next several months, I did not stay long at any firehouses. I was moved more often because I was there for just the engineer and when he came back I went to the next engineer vacancy. I was earning a little extra money but all the moving around was not worth it. The city was preparing for the Pan American games which were being hosted here. Firefighters were detailed to Spanish classes in the mornings and then a different group went after lunch, which meant I was being moved three times a day during the weekdays.

I started checking the vacancy list. One benefit of becoming an engineer was the ability to go regular (become a full time member of a crew and stay put instead of roving around) as soon as a position became available and a sub had enough seniority to get it. Since I had no seniority on anyone except Mary, I would have to get a spot where no one else would apply. That meant it was unlikely I would find a position as an engineer and would have to wait until my class went regular, picking a back step somewhere. Who knew when that would be. I kept a close eye on the vacancy list and applied for every engineer opening.

I was sent to the north side to drive an engine commanded by a nervous lieutenant. He was waiting for me by the apparatus, pacing back and forth, obviously aware of whom was coming. As soon as I arrived, he grabbed the C shift engineer from the kitchen, still holding his full, coffee cup, and brought him to me. I dumped all my gear on the ground so I could be briefed on the apparatus and not hold him up too long from being relieved.

The lieutenant paced continuously as the engineer told me about fuel levels, water levels, oil levels, and little quirks, such as putting the hook on the extended rod when it was in pump gear. Having almost burned up in the apartment fire when the engine jumped out of pump gear, I knew I wouldn't forget to place the hook in position. I was cautioned about the narrowness of the bay and how to line up the engine with certain markings so I would not take off a mirror or clip the firehouse with a bumper when backing in from runs. I could see the scrape marks along the bricks and support beams from other boo-boos firefighters had made backing up. I sure did not want to leave my mark there. I felt comfortable after going through everything twice for the lieutenant's benefit.

The engineer left me with a whisper, "He is always nervous like that. It is not you."

I got my housework done, which was to check out the engine, (already done with the other engineer) wash it and squeegee the water off the floors into the drains. It was better than cleaning the toilets which I had done for the past year. I was lost on the north side but was sure this officer would tell me where to go. I put away my gear and pulled out my jump rope to exercise. The lieutenant, the way he paced for hours, had a better cardiovascular session than I did. He was skinny, eating miniscule portions at lunch and dinner.

When the alert finally went off, I thought he would have a heart attack. He jumped into the cab, slammed the door shut, then grabbed the dash and door handle with both hands so hard his knuckles turned white. He was hanging on for dear life. He started giving me directions before I even hit the remote to open the door. He was a good officer who definitely knew his district. He repeated the directions twice before I turned onto the street after answering affirmatively to lights on, seat belts on, parking brake off and everything is a go. He instructed me where to turn right ahead, not yet, almost there, turn right-now; slow down; there is a stop light ahead (it was green).

He did not relax after returning from the run. He was as jumpy as a Vietnam veteran on his last day in the jungle, afraid he would not make it home. He knew the worst part was about to come, backing in. Making me stop in the street, he had everybody get out to guide me as I backed in, which is what was supposed to be done, but nobody ever did. I saw him in my side mirrors running back and forth with his hands on his head, pulling his curly hair out. He began jumping up and down as the mirrors became even with the building. He motioned me to stop and ran to my window. He was breathless from running around but managed to squeak out that I was about to perform the tricky part and to be very careful. I crept backwards into the correct position and put her in park, setting the brake.

I wondered if he slept that night. Possibly, he got a little sleep in between crash nightmares. The first brick hitting his head, after I crashed into the firehouse and collapsed the building, must have startled him awake anytime he fell asleep. He looked tired and a mess in the morning with his curly hair frizzed and the little bald spot showing where he ripped his hair out.

The next station had a grumpy lieutenant who assigned me to clean the stairs in addition to my regular engineer duties. The backstepper warned me to make sure I did not leave any fuzz balls on the stairs or the officer would go ballistic. I carefully swept and mopped.

Our first run was to a doctor's office. I asked which way to go and he said there was a real problem with the fire department when a young sub was made an engineer and did not know which way to turn out of the firehouse. At the end of the approach, I had to put on the brakes and stop until he reluctantly told me to turn left.

The doctor's office was close by. When we entered the office, we did not see a patient, only the doctor sitting behind his desk, talking on the phone. He pointed to the corner, interrupting his conversation to say only, the patient was behind the door. His patient had fallen out of the chair into the corner and was crumpled up back there. Grabbing his arms, I dragged him into the room where we had space to administer CPR. He was a big man and I was pushing hard to compress his chest. We had three on the engine, all busy with the patient. We could have used some help but the doctor never budged from his comfortable, chair on wheels. When the ambulance arrived, we struggled, because of the patient's size, to get him on the cot.

On the way to the hospital, I stayed in the ambulance doing chest compressions and the officer drove. The backstepper followed, driving the engine. The ambulance crew notified the hospital that we were bringing in a pulseless, apneic (no heartbeat and not breathing) patient. Our estimated time of arrival was five minutes. We rushed him into the ER still doing CPR as I was hanging off the side of the gurney. A waiting hospital team began working on him as we left.

I do not like to know the outcome of my patients. The statistics are grim; few come back after cardiac arrest and I would rather think they might have made it than know the reality which is most likely death. In all these years, I have not been able to understand that doctor sitting on his ass, not helping. He is a compelling reason why we need malpractice lawyers.

Getting into my shoes at night for first aid runs proved to be a problem. The shoes we were issued had to be tied and untied. In houses where they rushed to get out the door, I would slow them down trying to shuffle out in untied shoes. Some of the guys had slip-on shoes. I liked the slip-on idea because the crushed down backs of tied shoes which some guys wore, were terribly uncomfortable I wondered how they could stand to walk in them with the backs under their heels. Some guys cut out the backs after they got too crushed down. I picked up a pair of moccasins to slip on for night runs rather than crush the backs of my good shoes.

When we were dispatched at two in the morning on a stabbing, I was first to get on the engine because I was able to get into my new, slip-on moccasins quickly. We arrived to a room full of seated people, smoking and drinking. The police had not arrived but we went in anyway. Today we wait for the police to secure the scene before entering to prevent attack or injury, but back then we would stand to the side of the door, knock, then enter. We stood to the side because a couple of the old timers were shot at through the door.

We looked around but did not see any stabbing victim. After we inquired, one of the men pointed to a woman sitting on a stool. He said his woman stabbed her for flirting with him. As I approached her to get her blood pressure, she toppled off the stool into my arms. I laid her on the floor on her back. When I withdrew my gloved hands, they were covered in blood. The ambulance and police arrived in time to see my bloody gloves. The police pursued the perpetrator who, after the stabbing, ran out the back door with the bloody knife. We rode along with the ambulance crew, assisting the medic, who delivered a very drunk, low on blood, but living patient to the emergency room.

When we got back to the firehouse, the officer stopped me as I headed to the bathroom to pee. He said my shoes were not appropriate to wear on runs. They had soft soles so I could step on glass, or blood could soak through, which it had. Elsewhere, the guys wore sandals and flip-flops. Wearing my moccasins on night runs at the next house, no one said a thing. It all depends on your officer. It's almost like 90 little fire departments, (30 houses times three shifts) because each house on each shift is different.

Chapter 15

Frequently in the evenings, I called home from the firehouse and got no answer. Calling Mom, she said my husband phoned to say he would be late again. After he got there, he would jog before taking our baby home. My mom was getting perturbed due to the increasing earliness of his arrival and the frequency of his lateness. I was going to have to talk to him.

The baby was in his booster chair, having a spaghetti snack. My husband came in the side door from the carport. Walking over to kiss him, he had a smell on his face that I recognized instantly as I put my lips close to his.

I shivered and screamed, "What is that smell on your face?"

He backed away, his knees almost buckling (or were my knees buckling), his face turning white, and did not answer my frantic question.

I shouted again, "What is that smell on your face?"

I answered my own shouts, "There is only one smell like that in the world! You've been fucking around!"

Thinking of the baby, I didn't knock his head off or scratch his eyes out. For some insane reason, I started laughing as he denied everything. Did he believe I was that stupid? I wanted to know if he had been having oral sex with the woman he talked often about, a swinger, who got pregnant and then her husband left her for their swinging partner. He felt so sorry for her.

Reverting to the vulgar, a coping mechanism, I yelled, "Did you feel sorry enough to try and make her feel better by eating her pussy?"

I wanted to know but he denied it all. I felt like an idiot, having flagrantly disregarded the warning signs: his constant late nights, the infrequency of our lovemaking, his total distance. I felt like a fool for trusting him when I knew instinctively something was wrong. It did not occur to me that he would do such a thing. He had talked about her frequently but I thought it was merely talk. He had even introduced us at an ROTC function. Since it is such a distinctive smell, he could deny it from now until doomsday. I knew what he did and I was devastated. We fought a lot the next day and I reported for duty the day after. I was quiet at the firehouse.

The summer was over and I was not driving much. I had been stationed at this firehouse several weeks already. Not my jovial self, the engineer that I helped as he did the cooking, asked me what was wrong, while we peeled potatoes. I stammered that I caught my husband cheating. He said he had done the same thing to his wife and got caught but they made it through the painful situation and were still married. I wondered if I could forgive my husband.

After dinner, I got a bad case of indigestion so the engineer gave me a dose of antacid from a bottle kept in the refrigerator. All firehouses have big bottles of antacid. Sometimes, between our work and our home lives, we could down a bottle a week.

This firehouse was busy with first aid runs. One came in while the engineer was in the shower. The officer told the backstepper to drive so we could get out fast. It was a shooting confirmed by the police. We pulled up to a bar with angry people standing outside in the parking lot. There was weeping, wailing, shouting and crying. One person met us in the street and said to hurry because his buddy had stopped breathing. I hopped off my side grabbing the tech box that was full of bandages, BP cuffs, and first aid supplies. The oxygen kit was on the other side. The driver got the report book.

The empty-handed officer ran to the man lying on the ground and started chest compressions. He said we needed the oxygen bag so we could give him breaths with the bag-valve-mask, but it was still on the engine.

A man behind me was shouting, "Do something for him! He's my brother!"

I knelt beside his head. His face was gray, with fixed pupils and unblinking, unmoving, glazed eyes. He had small holes in his face and all over his chest. There was not a bunch of blood pouring out of him and yet he appeared dead. What caused him to die? It did not seem probable that these small holes, peppered over him, could have killed him! I had never seen anyone blasted with a shot gun before. I heard a witness telling the police that the victim saw two guys stealing some stuff out of his pick-up truck. He ran outside yelling, followed by a friend. As the thieves sped off in a blue truck, they blasted him! He fell down on the spot and his friend ran back into the bar to call 911.

I came out of my thoughts as I heard the brother, restrained by the police, yelling in a hoarse voice, "I am going to kick your ass if you do not help him! I will kill you if he dies!"

My officer, who had been speaking to me, grabbed my shoulder to get my attention and told me to get the oxygen kit. I looked at how far away the engine was as I heard a string of obscenities come out of his brother's mouth all directed at me.

Feeling I had no other option, I bent over the patient's face, squeezed his nose shut and gave him mouth to mouth respirations. The first deep breath I gave him, I pushed in hard. Putting my mouth over his, I administered a second breath and he puked in my mouth. Turning my head, I puked by his brother's feet. As I gave him more breaths, the ambulance arrived to load him on a cot. The medic gave me a bag valve mask or BVM to breathe for him. I told the medic that we had a response from the patient; puking in my mouth. Was that a good sign? The medic said it was not good. The three p's: puking, peeing and pooping are the last things people do before they die!

We worked him in the ambulance, performing our balancing act in the back of the moving vehicle, doing CPR all the way to the hospital. The medics put a breathing tube down his throat and started IV's, pushing medicines into his veins and alerted the hospital. I continued to bag him or breathe for him with the BVM. We wheeled him out of the ambulance to the emergency room where the waiting team of doctors and nurses took over.

They asked the medic a bunch of technical questions. I stood watching, unable to take my eyes off the action, while a nurse was asking me if I was okay. Distractedly, I told her I was fine. I wondered why she was asking me if I was okay. Then the doctors ordered everyone to stop, declaring he was dead. I asked how that could be! Even though there were a lot of holes, they were small. He said the patient had been shot through the heart suffering grave, internal damage and with a chest cavity full of blood nothing could be done for him.

The doctor, as he left the room, told the nurse to get me cleaned up. She walked me down the hall to a private restroom.

As we passed some people, they stared at me sympathetically saying, “Oh my God, she is hurt!”

Looking in the mirror, I saw blood covering my mouth and shirt. I was pale from puking and riding in the back of the ambulance, getting car sick and tossed around with every corner we turned. I felt hot and sticky from the arduous work of CPR but was beginning to cool down and feel clammy in the air conditioning. My stomach was unsettled and I was afraid I might puke again especially after seeing my vampire-like appearance.

Taking some deep breaths, I lathered my face with disinfectant soap. The nurse said she would bring me some mouthwash if I could be left alone for a moment. I wanted some mouthwash and assured her I wouldn’t pass out. When she came back I gargled, then returned, visibly upset, to my engine driver, the backstepper. The victim, my age, partying with his family, was gunned down in front of everyone. All those people, crying and wailing, crowding around his dead body, shot so senselessly, affected me deeply. The doctors did not cut open his chest and squeeze his heart because they knew it was fruitless but I would have felt better if they had.

When we got back to the station, the engineer was waiting for us.

The guys said, "She needs a hug."

He held me for a minute and then I started to cry. They quickly went into the kitchen and we went into the watch room.

He brought me a double dose of antacid. I said, "I used to wonder why you could find antacids in the refrigerators of almost every firehouse. I found out why tonight."

I told him the events and started crying again. He kissed my eyes, and then my lips. He unfolded the couch and we lay on the bed together. I cried some more thinking about my husband cheating on me. I looked up at my firefighter and we kissed passionately, undressing each other under the covers, I found comfort in his arms and love making. The closeness of his body and the smell of his freshly, showered skin helped me sleep. I had not slept well the last couple of days and I really needed some rest before I compromised my health.

We were startled awake by someone banging on the glass. It was the lieutenant. The engineer went to the door and spoke quietly for a minute. He came back to bed and told me to get dressed and go into my own bed. It was getting light outside and our relief would be coming soon. I asked if the officer was mad. He said no, but the lieutenant was nosey and wanted to know all of the details. The engineer told him to mind his own business.

This was the beginning of a deadly spiral for me. I was going down, spinning out of control and I could not pull out. I started seeing the engineer on the side, even bringing my son with me. We slept together at the firehouse on duty days. The lieutenant did not like it but since he had brought many women to the firehouse to sleep in his private quarters with him, he had no room to talk. He was jealous though and trapped me one day in the back of the firehouse while I was working out. He wanted me to go out with him. I asked him what he thought his engineer would think about that. He said he didn't have to know. Not the least bit interested in him, I told him I would never go out with him and nicely asked him to leave me alone. He threatened to give me a bad review. The reviews were sitting on his desk waiting to be filled out. I told him that my reviews were all satisfactory up to that point and if he gave me a bad one, I would have him downtown explaining to the spit-shine chief why he was sexually harassing me and pressuring me to have sex with him by making threats. Believing I would do it, he left me alone but regarded me suspiciously.

I moved to the next firehouse but talked to my engineer everyday, meeting on our days off. My son came with me on our visits, so we just walked and talked, holding hands. One day he called me and asked if I had seen the newspaper. There was an article describing the arrest of the two men who shot my patient. They were being held in jail unable to afford bond, pending trial. I cried, relieved they had been caught and would be punished for gunning down a man with whom I felt a personal attachment, even though he was a stranger.

I arrived at the next firehouse in time for Halloween. A neighborhood parade was planned and the two apparatuses were scheduled to be the star attractions. This firehouse had an engine and a truck. The truck was not like any I had seen. It was old. The backsteppers rode on the very back, holding onto a big silver bar that ran across the top of it. It was not an aerial but a service truck that was smaller than the engine. It carried various sized ladders, stacked on top of each other. There were a variety of truck tools-axes, saws, pry bars, long pike poles, a bowling ball to throw down chimneys for creosote, flue fires, etc.-in the compartments. One of the guys made dark blue t-shirts with a diplodocus on the front and “IFD Truck” “The last of the dinosaurs” printed below. It was a well designed shirt and I was happy to be given one, wearing it for many years.

As a safety measure, long leather straps were installed and secured to the top rung. We were supposed to wrap these around ourselves before the truck took off. It appeared to me that it would not prevent us from being bounced off the back, only dragged after we were bounced off, so I did not strap it on. That made the other back-stepper happy. He didn't want to wear it either. He was the station cook and loved to work out every day.

For exercise, we ran together, did some weight lifting, and practiced defensive moves to ward off attackers (he is a black belt in Karate). He was an instructor at his master's DoJo, no longer required to pay to work out. He had a lot of trophies displayed at the firehouse from winning matches at competitions. When I was in the army and in college, I practiced Tae Kwon Do. I caught on fast and he was impressed. The lieutenant worked out with us too. It was one of the few houses where someone worked out with me. It's nice to have an exercise buddy. It makes the workout go by fast and seem fun.

Being the new guy, I was sent to the store with a grocery list for lunch and dinner. Since we had two full crews, I was told not to take a radio; if anything happened, they would take care of it. They were used to running with seven. This was not a busy house and they did not want to delay lunch by making the shopper take a bogus first aid run. They had their priorities.

I walked because the store was directly behind the firehouse. When I got back from the store, pushing the cart with a pile of groceries in it, the place was deserted. I checked the printer for the address but the paper had been torn off and taken with them. I called control to get the address and the coordinates. The coordinates are the east-west, north-south position of the location where you are headed. If you know the coordinates of where you are, it will give you a good idea of where you need to go. Control told me my station was on a working fire. Shit! I had been lollygagging around the store, looking at all of the prices, attempting to be an economical shopper and missed a working fire! I could have just screamed.

Leaving the groceries in the cart, I grabbed my gear from the floor where it had been thrown, shoved it in my car and took off. I got lost on a street that took a trick curve but recovered from that making it to the scene where the fire was out. They said I was barely gone when the run came in. They thought about getting me but when possible entrapment was reported, they had to get out fast. There was no entrapment but it was rolling when they charged up. Now it was out and I missed it. They sent me back to the firehouse to put away the groceries. Nobody wanted the milk to go bad; again, priorities.

I set the table but had not started cooking when the crews came back. The cook started his oriental lunch as soon as he hopped off the truck. I could smell the smoke from the fire on him and was glum. He told me to cheer up; I would get plenty of fires in my career. Relating funny stories about his firefighting experiences, we were laughing pretty hard. When he opened the package of fish to cook, I said it reminded me of the smell on my husband's face when I caught him cheating. We had to take a break we were laughing so hard. My laughter turned to tears. I thought it would be a good idea if I cut up the onions so no one else would know I was crying. He agreed.

While eating lunch, we discussed what kind of costumes we would wear for the Halloween parade. We were scheduled to be out of service so we would not have to worry about taking runs during the middle of the celebration. The neighborhood association brought us candy to throw to the kids along the route. We had four boxes, plenty of candy, so I helped myself to a couple of suckers. I am a candy-aholic, after all.

A Boy Scout troop came in the afternoon for a tour of the firehouse. One of the first questions was where our pole was. This station was a single story building with no pole. The scouts seemed disappointed because they wanted to slide down the pole. We explained that kids could not slide down the poles anyway because too many ankles get sprained or broken by seasoned firefighters. Other injuries happen too when an excited firefighter slides the pole too soon and lands on the other firefighter still sliding the pole, squishing him or her at the bottom.

We donned our gear. In my eagerness to impress the kids, I put an arm through my tank strap, yanking it off the side of the truck where it was held in higher-than-usual brackets. I clunked myself right in the head and it hurt like hell. One of the scouts said I was too short to do the job right, so I picked up his dad throwing him over my shoulders in a firefighter carry, swinging him around. He had no comment after that. Some of the other kids wanted to be picked up and swung around so we gave them all a good spin and got them dizzy. The one kid refused to be picked up. As they were leaving, I heard them discussing, "girls on the fire department".

When we went in to make our beds, (I made my bed later now to forestall pranks) the cook was done in less than a minute. I took my time, as I usually did, tucking in the corners in military fashion, fluffing my pillow and smoothing the sheets. The cook said he had never seen a fireman take so much care making a bed and was I that fastidious about everything else?

I said, "I am not a fireman, I am a firefighter and that makes all the difference."

My dad said "fireman" sounds stupid, descriptive of a person on fire. "Firefighter" is apposite, making better sense describing the work we do.

That evening, the truck engineer, who was heavy into real estate investments and owned several houses in the area, began race-walking in circles around the perimeter of the bay, muttering to himself! The other guys said he did that a lot and not to worry about it. Walking with him for a while, I told him of my interest in buying some rentals. He gave me some advice, peaking my interest in the possibility of starting a business on my day off.

Joining the rest as they sat outside on the benches, they told stories and waved to passing cars. One of the guys had a couple of sons that dropped by and sat with us. They were interested in meeting and speaking with the lady firefighter. They wanted to know the application process because they were considering applying when they turned 21. My engineer called but I did not talk to him very long because I was missing the party outside.

The next day, for the parade I wore a blonde wig, lots of make up and a mini-skirt. We threw most of the candy to the kids leaving us short for the trick-or-treaters later that night. We stopped and talked to people at the end of the parade route. Hundreds of kids, each boosted up by us, boarded the apparatuses to blow the horns and run the sirens.

A man approached me and pulled out a wad of bills from his pocket. He said I was beautiful and wanted to meet me later on. I said I did not get off work until the next day and I had to hurry home to my husband and baby. The cook came over, looked at the wad of bills, and said he would meet him after work. The guy put his money back in his pocket and walked away.

The cook, offended, said, “I guess I wasn’t his type!”

The cook gushed that it was a big wad of bills I turned down and wanted to know what my asking price would be, guessing a million? I said some people would never have enough money and others could have it for free. He wanted to know which one was he. I just smiled. We made an emergency run to the store to buy a few bags of candy. We didn’t want any of the trick or treat kids turned away on Halloween night.

The cook brought in a mirror the next day and sat in the locker room on the benches, acid etching a woman warrior with an axe, helmet, and breast plate armor. Underneath her was a flowing banner emblazoned with NOW. When I asked him about it, he said it was me and the National Organization for Women. He gave it to me as a gift after he framed it. He was a tremendous artist and I was flattered to be depicted as a firefighter-warrior!

While we were cooking dinner, he got a call from his wife and I heard him arguing with her, then slam down the phone. He said some guy was calling the house a lot and he thought his wife was having an affair. I said there was a lot of that going around. He stated he wanted to have an affair with me and after some hesitation I admitted I was already involved in one, therefore I could not possibly do anything with him. He said he was not upset that I was having an affair; he just wished it was with him. I wished it was with him but no way would I say so.

Late that night, he said he wanted to show me something in the hose tower. Back then the hose towers were used to hang our hoses when they were still made with rubber linings. After fires, if any hose was put back into the engine bed while still wet, the linings would rot and leak making the hose useless. The towers were built tall enough to accommodate our 50-foot sections that were draped over a huge round column to dry, leaving two sides hanging approximately 25 feet. Today, our hoses are lined with a material that doesn't rot. The hose towers are no longer needed.

We climbed up together past the dangling hoses hung there after the fire I missed. He proudly displayed a mattress he had dragged up there, made with sheets smooth and tucked, and topped off with a fluffy pillow. We fell into it, wrecking the sheets after he put on a condom. I wanted to laugh and tickle but we had to be quiet since anybody could climb up the ladder if they heard something. As we finished putting our uniforms back on, the lights came on and the alert sounded. We almost fell over each other getting down the tower ladder; standing nonchalantly as everyone else came pouring out of the bedrooms. It was for the engine only. We waved to them as they left. The truck lieutenant looked us up and down and asked what we were doing up so late.

"Just could not sleep, sir."

After he went back to bed, we climbed back up the hose tower for another round as the engine roared off.

I was tired as hell when I went home. My mom had convinced me that Alex should be in preschool. I enrolled him in an academy three days a week for three hours at a time. I dropped him off and went back home to sleep on the couch. The ringing of the telephone woke me. I did not answer it but rushed out the door to pick up my son because I was 45 minutes late picking him up. They put him in a chair by the door where he sat until I came. My poor boy sat there for an hour staring at the wall. He was so active that it must have seemed like torture to him. He was very glad to see me. He put his little arms around my neck as I bent over to pick him up. He must have thought I abandoned him.

Back at home, as we were eating lunch, I wondered what kind of mess I had gotten myself into. I was not touching my husband with a ten foot pole but I was having affairs with two men at different stations. Plus, I was being a terrible mother, late to pick up my son due to exhaustion from being up most of the night, playing around. If I was exhausted from working, that was forgivable. Something had to give. I called my engineer and told him I could not see him anymore. He was upset but was gracious not to ask why.

Given my marching orders to the next station via a Centrex call, I told the cook. He was very disappointed. We were having so much fun in the hose tower every night. In the morning at shift change, he cornered me in the bedroom for a good-bye kiss. I tried to push him away but he assured me nobody would come in. They were all in the kitchen and I could give him a quick kiss. Just as we locked lips, the door opened and then quickly closed again. It was a guy from the other shift arriving late. Word would unquestionably get around. I did not want that to happen because I was not sure if I wanted a divorce or to just keep punishing my husband. I was having a lot of fun punishing him.

The cook and I continued our affair for months. He would come over while Alex was in preschool. He parked at a church lot and walked to my house so no neighbor would see a car sitting in my driveway all morning. One morning my husband came back to get his wallet that he inadvertently left on the table. We heard the car drive up. Luckily, I had put the chain on the door as a precaution. He could not get in with his key. My cook jumped up, grabbed his clothes, and ran into the other bedroom to hide in the closet. He was naked with a condom still on. I threw on my robe and answered the door like my husband had just awakened me with his knocking. He asked me if I felt okay, saying I looked pale. Coolly, I said I had a busy night at the firehouse and my stomach was a little upset. Both statements were sort of true. (My stomach got upset when I heard his car drive up.)

After he left, I let the cook out of the closet and told him he better go home because I had all desire knocked out of me by fright. We had both seen the results of shootings after a spouse caught the partner cheating. Since my husband had a gun in the house, I thought it best my cook should leave before one of us ended up shot and the other dialing 911.

I was soon involved with another firefighter but we made arrangements to meet at a hotel. He was a substitute like me so we were not at a firehouse together and I was not going to bring anybody else home. I would take my chances running into fires but I did not want to be dispatched to hell by my husband if he caught me in bed with someone else. It was Christmas and I bought my sub-lover a watch. The cook and the engineer were still calling me. The engineer was begging me to see him again and the cook was asking me to marry him because he was leaving his wife.

Always in on the fun, Mandy and I went to union meetings every month. We drank our peach schnapps straight out of the bottle wrapped in a paper sack as we listened to the treasurer's report and committee reports. After the meetings we went to our favorite local bar that played 50s music. I wanted to dance and have a good time. I had to go home to my husband after the meetings so I could not stay out too late. Mandy did not have anyone to answer to. She could do as she pleased and sometimes she was pleased to go home with one of the men. We shared information about all of our adventures.

Chapter 16

I got orders to report to a single company on the north side. Even though there were only four firefighters at this house, it was a bigger house with a pole. It used to have eight firefighters until the department shut down the truck. The year before they closed the truck, it had 12 runs. One of the guys boasted he got paid a full years salary for one run a month. He stopped bragging when he had to find a new firehouse.

When I walked in the back door, the driver met me and said, “We were just talking about you. We decided we need something to liven up the place. We need Kathy Gillette, and here you are on our door step. What luck!”

My first thought was, “Crap, word is out. That big mouth who caught me kissing in the firehouse bedroom blabbed all over.”

Placing my gear on the engine, I was shocked at the outdated equipment. I thought it would be better as an exhibit in a museum. It had an open cab with the bar in the rear for the backsteppers to strap onto. The leather belts were missing, which I wouldn’t have used anyway, and would again be holding on for dear life. There was a silver bell on the officer’s side. The bell had a rope that went to the driver who clanged it as he drove to all the runs. It used to clear the way of traffic in the days of open vehicles but served only as a showpiece now. The engine had been retrofitted with sirens for present usage but the driver loved that bell and kissed it every morning when he arrived and clanged it for the kids and parents when they came in for tours.

The lieutenant planned to retire in less than a year and was serious about it. He was not taking any chances. He would not slide the pole. He walked slowly down the steps for runs. We sat outside after dinner and talked about the people walking by. The lieutenant's girlfriend came by as we were sitting on the bench. She had a Valentine's Day present for him. After they left, I remarked that I thought the lieutenant was married. Had not his wife called earlier? Yes, he was married but he had been seeing this girl for several years. He was supposed to move to Florida but was considering changing his plans because he did not want to leave her behind. I left the guys to answer some phone calls from my lovers and one from my husband asking me to hurry home in the morning because he had to be in class early.

When I came back, the driver was by himself because it was chilly outside. We talked for a while until he suggested we play billiards. This house had a pool table and a big screen TV. We racked the balls and even though I am only so-so at eight ball, I won! He wanted to play a more interesting game, strip pool. He said it's similar to strip poker and explained the simple rules. For every ball that went into a pocket, the loser had to remove an article of clothing. Since it was winter, and I was cold natured wearing a lot of clothes, including my flowered, long underwear, which were now tattle tale gray, I took the challenge. I had him naked in no time and was still toasty in my long underwear. I could not take my eyes off him. He was cute and a little excited, sporting the beginning of an erection. It suddenly seemed very warm in the room.

We got dressed, turning our shirts inside out, and went across the street to a bar, bringing along a station radio so we wouldn't miss a run. We were sitting at a table in the back of the bar, drinking whiskey and coke, when my substitute-firefighter-lover came in the door. At his firehouse, he told them he had an emergency, a broken water heater at home that was leaking everywhere, and he needed to get away. He sat with us for about an hour. I excused myself to go to the bathroom, leaving the two of them to talk. When I came back, my sub said he had to get back to his firehouse, leaving with a friendly good-bye. The driver and I had another drink and left too.

In a giddy mood and feeling no pain, I agreed to play more strip pool. Being half drunk, I was the one who had to strip naked. He was a little drunk too, so he took his clothes off even though he won. He picked me up, placing his warm, soft lips on my breast, tickling my nipple with his tongue. Laying me on the couch, he cupped my other breast, and with his fingers tickled my nipple. The double pleasure shot through my body. He was fully erect and I was hot for him when he plunged inside of me. As we silently made love, erotically he stared longingly at me, never taking his eyes off me as he moved up and down causing me to tingle to my toes.

Embarrassed afterward because I lost all control, as I lay wrapped in his arms still naked, I started to confess my problems but he hushed me saying he already knew and did not care because he was in love with me. He told me that my substitute showed him the watch and asked him to leave me alone. Experiencing a luscious trepidation, I thought I might be falling in love with him. It was just too crazy. We were both married and had kids. The sex was too good to pass up, though. I was at this station because of a vacancy and was scheduled to stay for several months. We made love in every room of that firehouse including the bathrooms and the shower.

It was customary for the engineer and least senior person to take the apparatus to the shop for repairs. He drove while holding my hand under the dash. The wind was blowing through our hair in our convertible fire engine making us feel free. He took my hand and placed it on his pants which were bulging. As he drove, I rubbed his crotch through the cloth. Unzipping the pants, I pulled out his erect penis. Moving closer to him, I stroked it and caressed it. Gently he pushed me towards it. I closed my warm lips around the tip and ran my tongue around the head, sucking on the shaft until he climaxed. He almost ran off the road.

One morning, when he was leaving, my driver backed into my car, damaging the fender. He said I should follow him home to meet his wife and kids. We could show her the damage and exchange insurance information. All of his information was at his house. I followed him home. His kids showed me their hermit crabs and his wife was cordial giving me a tour of their house.

One night I was sitting on his lap when his wife called. She nagged at him about some cabinets that he never put up. He set the phone down, without hanging up, went to the bathroom and when he came back, she was still rambling on. He placed the phone next to my ear. I could hear her ranting. Screaming into the phone, she was calling him stupid and lazy. No wonder he was unhappy.

He brought me presents every day, cards and candy, little, chocolate, Easter eggs. When the Centrex rang for me to go to the next station, he was devastated. That night he bent down on his knee and asked me to marry him. I was not sure what I wanted. He said I should not play with his heart but I could not give him an answer.

Chapter 17

The next station had a psychotic weirdo for a lieutenant. They were in the middle of spring cleaning and he got upset with me because I wanted to exercise before we started cleaning. Even though there was a general order stating we were supposed to work out between 0800 and 1100, nobody respected it or followed it. He said I needed to start cleaning now. I could do P.T. in uniform later.

A couple of guys from the other shift stopped by informing the lieutenant that a "nigger" had put in for the engineer vacancy. I think they knew he was racist and were ragging on him. He took the bait saying he would run him off because he would not have any "niggers" in his house. I thought my reputation as a member of NAACP and NOW had preceded me but either these guys had not heard or did not care. Maybe they were setting him up. I kept my mouth shut and continued washing the walls. I was already having enough trouble with this nasty lieutenant. Thank heavens, the position would be filled soon by the poor guy they were talking about and I would be off to the next house.

My driver was having problems with his wife. She became suspicious since he brought me home. Her nagging increased. He wanted me to leave my husband and run off with him. We met on our days off. He had his son with him and I had my son with me, neither of whom were school age. We kissed in front of the boys. Sometimes we would do quick pecks and the boys would laugh. One time we hit a bump while kissing and just about busted our lips. We all started laughing.

When my driver was at the firehouse his son laughed at two people kissing on TV. His mom wanted to know why that was funny. He said it is funny when people kiss like Daddy and Kathy. She got details from her son and called my driver at the firehouse. Using the insurance information I left, she called my husband and gave him details. My driver called me and warned me. My furious husband called me. It was late and the lieutenant had been in bed for hours. He was not happy about the late night phone calls and told me so in the morning.

My husband had school in the morning but he saw me as soon as class was over. He told my mother about my affair. She was upset with me. I explained what started it all, Richard stepping out on me. She was livid with both of us for not thinking about Alex. I was incensed with my husband, airing our dirty laundry to my family. I called a friend and made arrangements to move in with her. I packed some clothes and took Alex with me. I stopped at my lawyer's office and filed for a divorce. My driver was kicked out of his house. I had no way of contacting him. Cell phones were not around yet. I talked to him the next work day. He said his dad let him sleep at one of his rentals. At least he did not have to sleep in his car like some other firefighters I had heard about. One firefighter going through a divorce was reportedly living in a camper in the parking lot behind the firehouse. He came in the station to use the bathroom and shower.

My husband called and pleaded with me to come home. When I swore I would not until he moved out of the house, he hired a lawyer for his side of the divorce. I was sure I wanted a divorce but I was not sure I wanted to jump into another marriage. My driver pleaded with me to marry him but I still couldn't give him an answer. The uncertainty was driving my driver crazy.

Mandy needed a day off. The fire department needs a body in a certain slot. It does not matter if it is yours or some other firefighter's so we can have someone work in our place just about anytime, as long as we fill out the paperwork and do not trade more than three days in a row. Mandy and I traded time frequently. Replacing her on the C shift, I was on my way to the shower, having finished my workout, when I heard the guys talking about a firefighter who had locked himself in his house and committed suicide. Control was dispatching the chaplain and some chiefs to the location. My driver had already called me several times, in distress, because he could not see me since I was stuck at the firehouse and because I had told him I did not want to get married right away. When I heard them talking, I gasped, afraid they might be talking about my driver, and asked for the address. They told me the address, which was his, and asked me if I knew what was going on. I swore I did not know anything. It was obvious they did not believe me.

Slinking off to a deserted hallway, knowing in my heart that he could not have committed suicide, I called his house. I breathed again when he picked up the phone explaining that his wife had slashed his tires so he got his gun and shot it, putting a bullet in the ceiling. His wife tried to get in the house but the doors were barricaded shut. She called the police and the fire department saying she heard gunshots and her husband might have committed suicide. He told me he was too embarrassed at this point to go outside in front of everyone. I told him to go outside and be a little embarrassed rather than wait until later because it would only get more embarrassing if things escalated. He agreed and went outside. His wife was outside telling everyone it was my fault. He told her to shut up, that it was her fault.

After that dramatic event, word spread around the fire department like wildfire. Little old lady gossips can't begin to compete with firefighter gossips. Every station I went to, the guys were hoping for and pestering me for sex. Some of the women firefighters declared I was a disappointment. My classmate said she hated going to firehouses where I had been because the guys were eyeing her thinking she might be like me. Mandy defended me.

The next firehouse, I was gleefully greeted at the door, "We hear you are getting a divorce, Gillette. Is it true a firefighter was going to kill himself over you? Did you strip naked and dance on the table for the guys at the last firehouse and give them all blow jobs?"

Sometimes stories passing through the department grapevine could become sensationalized. My divorce filing was published in the vital statistics of the newspaper just after the incident. What bad timing. I forgot that was the first page most firefighters read so they knew without me having to say a word.

The stress and lack of sleep was taking a toll on my body. I could not see my driver because, until he divorced, he was considered a married man. It did not matter that he was separated from his wife. The courts frowned on anyone who dated a married person and could deny me custody of my son due to immoral behavior. My spit-shine chief said my driver's wife was repeatedly calling headquarters in an attempt to get me fired. The chief said the fire department considered the situation a personal matter and it would be very unlikely for anything to happen to me. On runs where people cried over hurt or dead, loved ones, I cried with them. The officers were sending me to the apparatus to sit and wait rather than have me blubbering and making things worse. It was very unprofessional but I could not stop the tears.

I had some vacation time coming. I took a day to sneak up to my driver's firehouse. I sat on his lap hugging him. He said he had been very lonely without me and his kids. His wife refused to let him see his kids. He wrapped his arms around me real tight and held my hands. He told me he had sex with his wife but it did not mean anything. He wanted to marry me but he needed some comfort. He was squeezing me because he thought I was going to knock the shit out of him. I did not want to hit him. I wanted to cry. My divorce was almost final and he had not filed.

Visiting the doctor because I could not function, she put me on a mild, non-addictive, anti-depressant that would not affect my work. I could go on runs now without crying and I could sleep too. My divorce was finalized and my driver went back to his wife saying he could not give up his kids. I moved back into my house, promising to give my ex-husband half the equity in it when Alex graduated from high school or if I remarried or lived with someone. (I paid him a few years later.) I traded time. Bringing Alex, I went on a much needed vacation to Florida and stayed with my sister for a week. Relaxing on the beach, listening to the ocean, while Alex played in the surf and sand, aided my recovery.

Mandy and I continued going to union meetings. I was very popular because all the guys hoped to get lucky. I just wanted to dance and have fun. Mandy was settling down a bit. She met a firefighter she was crazy about and was thinking about moving in with him, maybe even marriage. My driver went to the union meetings too. Even though he went back to his wife, he missed me and wanted to see how I was doing. I drank most of the bottle of peach schnapps Mandy brought and could barely walk out the door. I dropped my keys trying to get them in the lock of my car door. Mandy, grabbing them off the ground, started to get in the car. My driver wanted to drive me home and he asked Mandy for the keys. They fought over the keys and who would take me home.

She finally threw them at him and said, "You'd better not desert her this time. Once is enough."

My driver and I went for a walk in the neighborhood by the union hall. I wanted to take a break so we sat down by the sidewalk in some grass. We were lying back, looking at the stars, which were spinning, making my stomach queasy. We both fell asleep. I woke up with some guy's face close to mine.

To a friend, he sized me up, "She's just drunk." As he kicked my driver, he said, "Get out of here. This is my mom's house and she doesn't want drunks laying around in her yard."

My driver came up swinging. I was trying to stand. He grabbed me by my waist, propping me up against him. The two guys were swinging on each other with me in the middle, nothing but a hindrance in my driver's left arm. The police arrived as I puked, getting it on the officer's shiny shoes. When the police found out we were firefighters, they told us to get out of there and not come back. Still too drunk to drive, my driver brought me home to an empty house. Alex was at his dad's. I wanted him to stay but he left, despite my pleas, saying it would be better not to start anything again.

Waking in the morning, I had a king-size headache compounded by a bad, stomach ache. It was time to pick up Alex from his dad. When I went out to get in my car, there was no car. I had to think where I left it. I swore I would never, ever get that drunk again. After vaguely remembering the events, I felt lucky I did not wake up in the slammer. I would have never heard the end of it if my mother had to bail me out of jail.

At every station where I subbed, I heard disturbing rumors about the African-American engineer with the psychotic lieutenant who swore to run him off. It was conjectured he would be busted to private for incompetence. Realizing I knew him and had been driven by him a time or two, he seemed like an okay driver to me. I was certain the charges were trumped up so I called to let him know what was said before his arrival. He asked me if I would go downtown and report it. I told him he needed to report it, then call me and others as witnesses to confirm what the lieutenant said. I told him who else heard the heinous remarks so he would know who his other witnesses should be.

A few weeks later I was called by the officer, a lieutenant, representing my friend at the hearing. I wanted to know if anyone else was testifying. He said yes.

I was warned that "the whole thing is bullshit, an officer being dragged down by a black and a woman". I believed, at first, most firefighters would be disgusted with this creepy lieutenant's behavior. To the contrary, many seemed to be supporting him.

Calling the engineer's lieutenant days before the hearing, I expressed my anxiety about the situation. He visited me at my firehouse and we talked outside as we sat on a park bench. I was worried my life on the department would become miserable. I was getting cold feet, feeling the contempt at the firehouses. I told him that I would enter a room and talking would cease. The guys would act engrossed in their TV show. As I left, the conversation would resume. I wanted reassurance that my testimony was important because if it wasn't, I would rather not be at the hearing. He said my testimony was vital and they didn't have much of a case without it. Weighing his words, I decided there was no other way except the present course. I braced for the backlash.

At the hearing, the officer representing the engineer accompanied me into the room. The chief of the department and his second in command were sitting behind a desk. A union representative was in the room. The creepy lieutenant was not. I told them exactly what I heard. The creepy lieutenant claimed, the union rep said, I was doing this for revenge because he had given me a bad evaluation. I explained it was just one of many evaluations and it had no effect on me because the others were all good. It was not in my permanent file because it had been tossed out by my chief who did not believe it.

As the lieutenant and I walked out of the hearing, he shook my hand saying I had done a brave deed. In my car on the way home, I cried because I knew I would be shunned by people I thought were my friends.

A few others must have corroborated my testimony because the administration handed down a strong sentence. The engineer's lieutenant called with the results. Everyone on that shift was removed from the house except the engineer. They all received fines due to their complicity. The creepy lieutenant was given 30 duty days suspension (the equivalent of three months). After he returned to duty, he was placed in a double company for six months where he was to be monitored by a captain and was never to return to the A shift.

My brother filled the vacancy on the back step and remained there with the engineer until they shut the house down a few years ago. In our opinions, the engineer's, my brother's and mine, the racist lieutenant was a probationary officer and should have been demoted, but the decision makers didn't ask our opinions.

At the next union meeting I attended, when I sat down, everyone moved away. It was disheartening because I thought they were my buddies. After the meeting, at the nightclub, I cleared the table there too except one guy who said they were all idiots and I should be proud for standing up for what was right. I was surprised when another firefighter asked me to dance. I guess two people weren't mad at me. When we got out on the floor, out of earshot of everyone else, he called me a "fucking cunt".

I stopped dancing, looked at him and said, "Your friend was wrong to treat another firefighter so badly just because of the color of his skin."

I walked away. Only in rare instances did I attend union meetings after that. Our union serves all firefighters well by negotiating our raises and benefits. It has a problem though. It is dominated and controlled by white males who don't embrace its minorities. Subtle racism has caused friction with the minorities. Due to this exclusory behavior and other reasons, a separate organization was formed for black firefighters. Women need, but don't have, separate representation. The need becomes apparent when any male is disciplined due to a female's complaint. This is a problem with our international union as well.

Thankfully, all of the positions I had been applying for paid off. I nabbed a permanent position as an engineer on a north side truck. I would get a big, fat raise for driving and I would not have to endure the cold looks and silent treatment as I came through the doors of the firehouses. The rest of my class were still stuck subbing.

Part Three

Going Regular-A Home

Chapter 18

Going regular three months before the rest of my class, I became the first, female engineer in the history of the Indianapolis Fire Department but not a word was said about it. My two years as a substitute firefighter were over. I had my own house now. The station was racially mixed and I was received warmly. In honor of my arrival, they painted the identifying numbers on the tools in pink. I cut back the anti-depressant to half the dosage. After six months of being regular, I quit taking it altogether. The best time of my career had begun.

My assignment was on a 100-foot aerial. There are approximately three engines for each aerial. Aerials cover a bigger area for fires but the engines have more runs because of first aid (EMS), unless there is a squad. I believe there were six squads back then. Our engine crew went out at night a lot for EMS while, we, the truck crew, got to roll over in our beds. One of us had to get up and turn the radio and lights back off, usually me, since I was the least senior. I heard the engine backing in after their night runs, then I would fall back asleep. Sometimes, when they had a gruesome run, they would go in the kitchen and talk about it. I would join them and listen to their heart breaking stories. I was glad I was not serving on EMS runs. I definitely did not want to see any more shot gun victims or little, old, rigored ladies!

My boss, a captain in his early 60s was lanky and had thick, coke bottle, glasses, behind which he wore contacts. He was blind as a bat without his glasses and contacts. He had thicker glasses that he wore at night because he could not sleep in his contacts. He bumped into a lot of stuff in the night. He had a sense of humor almost as thick as his glasses. As we sat together for hours in the locker room painting our ceramic pieces, he would stretch intermittently saying it was hell to get old.

"Now, when I sit down for any length of time, I get stiff. The hell of it is that I get stiff all over except the one place I wish would get stiff."

He told story after story about all the guys on the job who did hilarious and outrageous things on fires. He entertained me endlessly.

The house had a truck bedroom, the quiet side, and an engine bedroom. The lieutenant on the engine snored really loud so I was glad he was in the other bedroom. Sometimes an engine crew member would sneak over to sleep in our spare bed if the lieutenant was having a particularly loud night. He kept a mini-refrigerator by his bed that his wife had given him for Christmas. If he woke up thirsty, he did not have to get out of bed to get a drink. He got fat from drinking all that soda pop in bed. As he expanded, he became the talk of the fire department, especially after he once got stuck between the seat of the engine and the siren.

My skinny captain made fun of the hefty lieutenant on a daily basis. The lieutenant claimed he could be a Chippendale dancer (a male stripper). The captain said the lieutenant would have to be a Clydesdale dancer because he was so big. The lieutenant was envied for his beautiful wife who made more money than the captain. The lieutenant and his wife bought a house that was practically a mansion. He threw a bunch of great parties but the captain, too jealous, never came to any of them. The lieutenant took it all in stride.

There was a vacancy on the back step of the truck. Sometimes we had subs fill the vacancy but most of the time, we rode with three. It was a common practice to run the trucks short in those days.

My backstepper was a wild character who loved to drink and party. He lived in a small apartment with his girlfriend who also liked to party. He was divorced and had his child support drawn directly from his pay. What was left over, he spent on booze. He drank some at the firehouse but not too much because the captain did not approve. He had a stash in his locker and would take a nip every once in a while. The captain and I put a crimp in his behavior because we sat together at a long, wooden table in the locker room painting ceramics. We would spend most of the afternoon painting. When we had enough ceramic pieces completed, the captain would take them home and fire them in the oven in his basement. I have many of those pieces still.

When I first arrived, the captain's daughter was getting married. He was making little, white swans for each place setting. I helped him paint the swans. He fired them and brought a couple of extras from the wedding back to me. They were stuffed with candy! Everyone realized quickly that I like candy and most everyday someone brought me some, spoiling me.

My little world was not perfect though. There were two other shifts and somebody did not like me. Some mornings I came in and found things hung on my locker like used condoms, dirty underwear, and other assorted nasties. I threw them away without making much of it. Some coward was afraid to confront me and was secretly trying to harass me. I considered the source and did not worry about it. I looked at it this way; if the others agreed with him, he wouldn't have had to be so secretive about it.

I was new to the north side and because of that, I frequently went the wrong way on runs. It became part of my daily ritual to spend 30 minutes studying a map. But looking at a map and driving there are two different things. I have a poor sense of direction which I did not realize until I started trying to learn the district. I seemed to find a lot of dead ends. There were times when I was absolutely sure I knew where I was going but ended up lost, lost, lost. The truck was a 100-foot aerial, not easy to turn around on a little side street. My three-point turns improved dramatically.

Every Friday and Saturday morning, I parked the truck on the approach and raised the ladder. I loved the challenge of raising it. All of the difficult and dangerous components of bringing a ladder out of its bed were right there on the approach of the firehouse: a tree and its limbs on the right; killer, overhead, electrical wires along the street. On Sundays we went through the truck tools, starting the saws and the generator that ran our fire scene lights. We cleaned out the compartments too. I washed it every day, driving it onto the approach so I could squeegee and mop the apparatus room floor. No more toilets for me, my house work was my apparatus.

I loved going fast. The engine engineer, in his new, turbo-charged engine, drove faster. It was one of the beauties we drove in engineer certification school. When we went on runs together, the engine pulled out first and I followed in the truck. Engines are generally faster because they are smaller and lighter than trucks. That order is not followed everywhere but that is the way the majority of houses work. We would speed down the road in tandem, jamming on the brakes when it was time to stop. On longer runs we came back and the wheels were smoking and stinking. If we put water on them, they sizzled and steamed. With all of this wear and tear on the brakes they were not stopping as well as they should. Every workday, I reported them to the shop (the fire department maintenance team) as needing repair.

One morning we were dispatched to a house fire with reported entrapment. The engine and truck arrived together. Flames were shooting out of the upper story and smoke billowed out of the first story. The engine company was busy pulling off the line so they could get water on the fire. My captain stepped out of the truck in his helmet, coat, gloves and boots. In five of his long strides he was in the house. Before I could finish getting on my gear, I was running up to the house with my SCBA half on and half off, the captain came out of the smoke, wheeling a woman in her wheelchair. She was coughing and sputtering but she was okay. She had become disoriented in the smoke and could not find her way out. The captain parked her and her wheelchair in the front yard and went back in to help the engine crew who were just now dragging their line in to attack the fire. Other companies started arriving. They stopped to talk to the little lady in the wheelchair. I followed the captain in. We had the fire out in a few minutes and started overhaul (cleaning the place of debris to ensure no burning embers are left).

After the incident, as we picked up hose, I told another captain at the scene how my captain rescued the lady. He said he would put my captain in for a commendation but never did. I tried to get the captain to write about his rescue so he could receive an award but he said it was just part of the job and the awards were political anyway. He swore politics was how he made captain. He had a neighbor who was heavy into politics and helped him get promoted. The chief sometimes bragged about how smart he was.

The captain told him, "You got your chief's badge the same way I got my captain's bars. You know somebody, only, your somebody has bigger nuts than my somebody." ("Having nuts" is how the guys refer to a politically connected person).

The chief was a screamer. No matter what you did on a fire, he screamed. Even though he was a pain in the ass with his loud mouth, the captain said he could be trusted to keep any problems in the house, and away from downtown. I did not like him because he questioned my ability to operate the aerial. One time we had a fire in a big warehouse with other attached buildings. The fire was determined too far along to go inside so we fought it defensively outside by the surround and drown method, aerials on all four sides of the building pouring water on the structure. I set up my aerial, which I did with ease since I practiced it two to three times a month, then I directed the water at the fire. The chief came by and sent me on an errand, replacing me with the lieutenant from the engine. The chief kept me busy doing errands most of the fire. I found out later, he did not want me operating the aerial because he did not think I could do it correctly. The lieutenant who drove him around, his aide, confided to me years later, that he never understood why the chief believed I was incompetent, but the chief definitely believed that.

The engine engineer was a good looking but shy, quiet type. His muscles rippled on his tall, well built body. He loved to watch westerns on TV and spent most afternoons in the kitchen glued to Gunsmoke and shows like that. He asked me to the state fair and we dated for months. The captain didn't approve but since we were both single people, I told him it was a private matter and we would not let it affect our performance at the firehouse. The captain let it be.

The engine engineer and I lifted weights together in the mornings. I would run around the track that belonged to the deaf school behind the firehouse and then take a quick shower so I could be in uniform before the chief arrived. He liked to give me a hard time if I was not dressed by his arrival. He never said anything to the ones who did not work out, but by God, I had better be in uniform on time. Selective enforcement is a hallmark of the Indianapolis Fire Department.

The engine had one backstepper who was there when I went regular. The other position on the back step was vacated frequently. Not many people could stand the lieutenant's snoring. I had a couple of guys tell me that if they beat him to bed and could fall asleep first, then they did not notice his snoring until a night run would come in. When they all went back to bed at the same time, it was a hopeless case to get back to sleep before he did. Therefore, we had a revolving door for the other back step.

We did not eat our meals together like most other stations. The engine engineer and I ate together. He would bring in big salads with everything on them, like tomatoes, eggs, bacon bits, broccoli, cauliflower, onions, etc. The captain and the backsteppers brought in leftovers from home. The lieutenant brought in a picnic basket of food and polished it off by himself everyday. The captain swore if you put a whole cow in front of the lieutenant, he would eat it. Not eating together may have been another reason why we had high backstep turnover. I always wondered if I was part of the reason. Perhaps no one wanted to come where there was a traitorous woman driving. That thought nagged at the back of my mind.

The engine backstepper loved to sing. He had a high, beautiful voice and dreamed of one day making it in the music business. He sang with a group that sometimes landed gigs at local nightclubs. He taped some of their music and we would listen to it in the locker room on his little, tape player. He had a wife and two kids. His kids were latch-key so they called the station when they got home from school, checking in frequently. The kids also called in the mornings before they went to school. He usually got the last call at night and the first call in the morning. He got a lot of calls. He was told to hang the phone around his neck.

After I begged for a basketball goal, they installed one out back in the parking lot. We would play sometimes and shoot around. Other stations came by for a volleyball challenge. We set up a net and killed each other. We had a lot of twisted ankles until a memo came out that forbade playing volleyball on duty. We obeyed the memo until the next spring when volleyball and basketball fever set in again.

We were a happy crew. The other two shifts complained about us not eating together. They said we caused the station commissary costs (each pay day we all pitch in a certain dollar amount to cover the cost of potatoes, onions, ketchup, butter, spices, etc.) to be higher because we ate a lot of peanut butter and jelly, and potatoes. It was standard fire department behavior to blame the other shifts for all the bad stuff that happens. We told the other shifts they made a lot of extra garbage for us to carry out front by eating so much. We used to put the garbage out back in cans that had to be hauled out front every week for garbage collection and then hauled the empty cans back. Thankfully, we now have dumpsters that the city garbage trucks empty.

When the regular officer or engineer is on vacation or sick, generally a backstepper moves up and fills the position. It takes a lot of concentration to move through the neighborhoods in those big rigs. Experience helps with all the things an engineer has to think about when driving. When a backstepper moves up to the position to drive for a few days, the pressure can be tremendous.

On the engine, the singer moved up to the driver's position. On a run, he was rumbling down a side street when a cat ran out in front of him. He jammed on the brakes but ran over the cat, squishing it flat. This greatly upset him but he couldn't have prevented it. Two runs later, a similar situation occurred and he ran over another cat. He was practically in tears when they returned from the run. Of course, sensing his doldrums over the situation, his teammates couldn't resist poking fun at him, hopefully getting him to laugh.

An announcement was made over the P.A. system that a tally was posted on the scoreboard in the kitchen. When we all piled into the kitchen, we saw the chalk board had his name on one side with a score of 2 and on the other side was "cats" with a score of 0. Throughout the day the score was announced over the P.A. system each time he returned from a run. The engine crew asked the truck crew if they could borrow a pry bar to get the dead cats out from between the dualies (the double set of tires on the back). A thump-thump sound was made over the P.A. system. The speaker said that was the sound the dualies made as they turned and the two wedged cats thumped-thumped against the road. The jokes multiplied as the day progressed. He had to laugh.

We got a new backstepper on the truck. He was the one who called me a "fucking cunt" for getting his racist, lieutenant friend in trouble. He apologized for his action and admitted he was wrong for calling me that and his friend was wrong for his prejudice. He said he was impressed with my professional manner throughout the entire situation. I felt a little vindication but was still uncomfortable with his presence. Although I didn't trust him, it was nice to have another person in the house who could share the load, especially on fires. He was exemplary at throwing ladders.

One day we were dispatched to a huge, fiercely burning, junkyard fire. No one was aware that magnesium rims were part of the junk. Magnesium is water reactive and can cause great explosions if water is poured on it to put out the fire. We arrived and were sent to a corner of the yard to set up our aerial and throw thousands of gallons of water on the fire. I did the usual elevate, rotate and extend. When I got it positioned, our new backstepper climbed up the aerial, locked in with the leather, safety belt, and directed the stream of water onto the fire. Other aerials were getting set up to do the same thing.

Suddenly, a big explosion boomed, rocking the aerial's extended ladder. My new backstepper was knocked off his perch and dangled by his leather safety tie-in. I yelled and started to race up the ladder but he regained his footing and shouted he was okay. My ears were ringing from the blast. A shower of molten debris came raining down on us. It made little melt spots on our helmets and coats.

Not having a permit, the owner tried to hide the problem. Finally he admitted what we were dealing with and the water supply was quickly shut down. The Hazardous Material Team was requested and dispatched. We waited for their arrival while the news choppers flew overhead. Every once in a while another small explosion would occur giving the news team something else to film. The Haz Mat Team radioed the chiefs that they did not have anything for magnesium fires but the brass, thinking it would look bad to the media if they had everybody waiting around doing nothing, ordered the Haz Mat Team, upon their arrival, to advance a line using Purple K. Even though I wasn't on the team, the lieutenant let me help move the line. The Purple K looked pretty cool coming out of the hose with its beautiful, purple mist but it is a water based chemical which caused another mini explosion that almost knocked those of us on the line over. We withdrew. At least the chiefs could say they tried.

We hung around the fire all day and night. We were relieved in the morning at the scene as it continued to burn for several more days. We spent the next workday, morning and afternoon, protecting exposures. Finally we were released and the companies filling in for us got to go home too. (Whenever there is a large fire with multiple companies dispatched to a scene, if they are expected to be there for an extended period of time, crews from other sides of town are sent to take the place of the dispatched companies. That way there is not a huge gap in service with companies having to travel great distances to a different emergency.) The poor engine crew got a run right after they marked back in service before they could even back into the bay.

The new backstepper wanted to be certified to drive the aerial as back up engineer. Being the regular engineer, naturally, I supervised his practice. The aerial ladder sticks out ten feet in front of the cab above the driver's head. We had a fat rope with a knot at the end dangling from the top rung of the ladder as a reminder- the ladder is sticking out here. Engineers have to turn aerials wide around corners so they do not run over sidewalks or anything else with the back wheels. At the same time, we have to watch the front so we do not take out anything on the opposite corner with the ladder tip. It was tough to make a tight corner and it had to be done slowly and carefully.

After driving a few times, the new backstepper gained some confidence. He was doing well until he took a corner too wide and wiped out a stop sign with the aerial tip. I saw what was about to happen and shouted but my reaction time added to his reaction time did not allow enough stopping time. He was thoroughly embarrassed by the incident and moved shortly thereafter. My cold countenance around him did not help. His little boo-boo was child's play compared to what I did. If he had hung around for what came next, he could have felt vindicated.

Chapter 19

In our district, we had a deaf school, college campus, and a high school that we visited frequently. Mostly, we had false alarms at those places. On these and other runs the brakes were taking longer to stop the aerial and I continued to turn them in every day.

One morning I came in and the brakes still had not been worked on. I was told another aerial in the area was having the same problem so they were in a reserve (an older apparatus used to replace vehicles until repairs can be made to the regular apparatus) while it got new brakes. Control did not want both aerials out at the same time. I advised the captain, we should mark it out of service for safety reasons. He said we couldn't refuse to drive it. He said I should just slow down. The speedometer was broken too. Budget issues are resolved much higher than my pay grade and orders are orders.

That afternoon, a run came in on a street that I didn't know. The engine engineer was on vacation. We had only a vague idea of the location. I knew by the coordinates that it was going to be a long run in heavy traffic. There were a lot of stoplights to go through, which meant heavy braking-just what I didn't want. I kept a safe distance between the engine and my aerial. I didn't want to get too far away from them because the traffic at intersections stops for the first emergency vehicle and if the wait is too long, the traffic moves into the intersection not seeing, hearing, or expecting the second emergency vehicle.

We went through several intersections, heating up the brakes. We were traveling on a divided road with two lanes on our side and two lanes of on-coming traffic. There were 6-foot-drop-offs on both sides of the road. Cars were pulling over like they should, and we made excellent progress. I suddenly realized the engine passed where it should have turned. The driver slammed on his brakes and started backing up. I put on my brakes, but when I saw their backup lights come on, I yelled to my captain to hold on. I was standing on the brakes with nowhere to go but straight ahead. The engine was cock-eyed in the road as it backed, taking up both lanes. Aerials have very little between the driver-me in this case-and what ever is ahead. I was going to smash right into the engine and kill my captain and me because the brakes weren't stopping us. I screamed. Aerials weigh almost twice what an engine weighs. I prayed I wouldn't kill the engine crew too.

My ladder hit the top of the engine, tearing up the hose bed and knocking their portable ladders 50-feet into the air, flipping them end over end into the yards of the houses on the side of the road. The aerial cab smashed into the back of the engine, caving-in at the dash, collapsing the steering wheel into my lap and pinning me. My officer was pitched forward into the windshield. I couldn't see what was happening to my backstepper. As the aerial moved forward, the engine started crumpling. The huge aerial squeezed it like an accordion clear up to the pump panel, which is just behind the backsteppers.

The aerial, having spent its energy, halted. The captain and I were alive. I wiggled out from behind the steering wheel. My hip hurt, but I didn't think it was broken. The captain, searching for his glasses, had a headache but said he was okay. My backstepper was totally unharmed. The guys were jumping off the engine, coming back to see if we were okay. The singing backstepper was holding his back but said he thought it would be okay. The other backstepper was a sub and said he was okay. Everybody was shaken, but okay.

The captain got on the radio and reported the accident. Everyone in the city fire Department heard it as the captain reported it. Control started other engine companies for the run which turned out to be a false alarm. The fire department sent a chief and a lieutenant to take pictures. Ambulances were dispatched to bring us all to the hospital. The chief, after surveying the damage, said we all needed to go. He marveled that we were still alive, much less virtually unscathed.

The media monitors the police and fire channels. Reporters rushed to the scene and the hospitals. The fire department, based on past history, knew we would be mobbed so they snuck us into the hospital through a back entrance. Our firehouse was shut down until fill-in companies could be dispatched. At the scene, I cried as I inspected the damage, but had calmed down by the time we arrived at the hospital. Even though I felt sick to my stomach, I managed to keep from puking.

We were all examined. The nurse told me I had the lowest blood pressure of everyone. It wasn't surprising, since I was the only one who worked out every day and had already released the tension by crying. My headache, probably from rocketing blood pressure and shots of adrenaline at impact, was subsiding. I was left with a bruised hip bone, but no permanent damage. We were all released back to duty, except the singing backstepper. His back was hurting. He took a couple of days off.

We were assigned a reserve aerial and reserve engine at the shops. We took our equipment off the damaged vehicles, which had been towed to the shop while we were in the hospital. Before midnight, we were back at the firehouse relieving the companies who filled in for us. A mechanic at the shops told us the engine was specifically designed to crumple like it had which probably saved our lives. If it had been an older engine, he was certain the captain and I would have been crushed. He wondered how I wasn't chopped in half by the steering wheel which was pressed into the seat. He said it was a good thing I wasn't equipped with certain male appendages!

Tossing and turning after I lay down, I couldn't sleep. Sitting up on the edge of the bed with my head hung low, I forced myself to stand and walk into the kitchen. The captain was drinking coffee. We sat through the night, both of us bleary eyed as the other shift arrived. We had to tell them what I had done. I wanted to run, hide, and never show my face again when I saw their stunned reaction. Nobody yelled or pointed fingers. Mostly, they consoled me by patting me on the back.

Driving to my mom's house, I went in to pick up my son. Now that his dad had moved away, Mom and I were his only caretakers. Crying again, I related the sad event. She said she heard about the wreck on the news and was worried it might have been me. "Well, if they had fixed the damn thing, like you asked, it probably wouldn't have happened." Good old Mom.

Headquarters sent a memorandum, delivered by the chief that a thorough investigation of the crash would take place. A specialist was hired to examine the brakes. Most of my co-workers wrote letters about the problems they experienced with the brakes. My backstepper testified that he felt the brakes applied in plenty of time and that they faded. The captain testified that I was a good engineer. Even the grumpy chief reported the many times I had asked for the brakes to be repaired. He had noted each complaint in his log book and submitted it for evidence. Their own investigation was inconclusive. The specialist said the brakes did not totally fail, but were greatly reduced in effectiveness due to prior overheating and glazing. Due to worry, my mouth was covered in cold sores and my face broke out in zits like a teenager's. At the hearing, I lisped through my thick, swollen lips.

Word leaked from headquarters that I was going to be stripped of my engineer certification, but with the room full of people pulling for me, the divided board decided to let me off with a year's probation. Nowhere in the process, not even at the accident scene, was anyone ever ordered to take drug or alcohol tests. Never did I have to talk to the media. The entire incident was kept internal. I thank the city and the fire department for that. I was pardoned, but not forgiven. The incident was talked about for years, earning me the nickname "Crash". To make light of it, the guys printed a t-shirt with a comical depiction of the crash and gave it to me with a piece of the mangled bumper. The running joke was my captain didn't even realize there was a crash. When I reached for the shifter to downshift and slow us down, the joke went; I grabbed his dick by accident. He thought, when his glasses flew off his face, it was only his climax.

The captain didn't think it was funny. Neither did I. We just had to grin and bear it until something new came along for our colleagues to gossip and joke about. On the fire department, that doesn't take very long.

No one testified the engine was backing up. Many firefighters on the department will probably be surprised to read the engine blew by the street, slammed on their brakes and backed up whoppy-jawed in the middle of the street, contributing to the accident. The sub backstepper present that day, asked me years later why I never said anything. I told him I didn't want anybody else to suffer for something that was mainly my fault.

Chapter 20

My screamer chief turned 65 and decided to retire. The chief who welcomed me my first day in the companies, when I was all wet, and whom I beat at poker, moved to our battalion. My captain didn't like or trust this new chief. A lot of people in the department felt the same way. This chief had a military background where they would rather court martial someone than stick up for them. I understood that attitude, because I had served in the military and was glad to be done with it. Even though the fire department is considered a paramilitary organization, (we can't strike because we are sworn to uphold our duty) it is nothing like the military. There is fraternization among the officers and troops. We live, eat, play and fight fire together.

Sometimes the people, who are downtown working nine to five through the week, get a mighty-high attitude because they forget what it is like in the companies. When politics change and they get booted out of their jobs, having to crawl back to the companies, some of them remember what they missed, the camaraderie.

One day, the afternoon peace and quiet was shattered. An alarm for a residence fire was sounded and the house was in the heart of our district. Arriving quickly we saw heavy smoke showing in a double residence. We entered the side of the residence where the fire was reported. The house was dark and full of smoke with almost no visibility. The captain and I were asked to set up lights from our generator so the engine crew could see their way around the house without falling over stuff. I headed out the front door as they ascended the steps. Starting the generator and hooking up the cord reels, I went back with my arms loaded in reels, lights, and cords.

As I bent over in the doorway to plug everything together, an explosion ripped through the house. Debris, hot air, and splinters flew past me. I kept my head down, bracing myself in the doorway. It was over quickly. My ears felt clogged from the concussion. My hands were stinging from the splinters that had been forced into them. (I had removed my gloves to plug in the lights.) Fear turned my stomach upside down as I thought of the firefighters deep in the house. Becoming careless, I had not donned my SCBA. I ran back to the truck to grab it and put it on as I ran to the house. Injured firefighters were tumbling out the door. I helped them walk out and sat them down in the grass.

Other companies arrived. They hooked up and attacked the fire in the house which was now fully involved. The engine backstepper, the one who liked to sing, came out and sat down on the grass. He said he was hurting. As I helped him take his SCBA mask off, half of the skin on his face came with it. My shocked reaction startled him. Taking his coat off, the bad burns on his wrists became apparent. His hands were burned too. Everybody was burned except the captain and me. He was also outside when the explosion occurred.

The captain and I helped load the injured into ambulances. After the doors were shut, and they took off for the hospital, I grabbed an axe and started chopping walls in the house. Several of the guys noticed my insane chopping. One came over and grabbed me, taking me outside. He sat me on the back of an apparatus, while a captain, acting as safety officer, talked to me. They both assured me that my fellow firefighters were going to be okay even though it seemed grim. The safety officer wanted to make sure I was okay. He took off my gloves and wiped the blood off, picking out the remaining splinters. As I wiped off my face with a shaky hand, a photographer snapped a picture of me. I wanted to get back to work and didn't want any more pictures taken of me. The captain released me to chop the hell out of the house.

In our post incident analysis, we discussed what went wrong. The side of the double we entered was not the source of the fire. It was in the attic on the other side which had vented itself and was free burning. A free burning fire doesn't create much smoke. Our side, on the other hand, was heating up, building up smoke, without enough oxygen to burn freely since the fire on the other side used the oxygen. When my engine company, at the top of the stairs, opened the door where the smoke was coming out, they introduced a rush of oxygen into the overheated room, causing a backdraft. The explosion of flames surrounded them with nothing but orange and shot them backward down the stairwell. The big lieutenant landed on a female substitute and broke her foot. The singer, who was at the front of the line and opened the door, landed on top of everybody so he didn't break anything but because the backdraft washed over him, he got burnt the worst. Another guy who bailed out a window also got some bad burns, losing part of his ear. I have always felt bad that I wasn't in there with them.

Once again, my engine house was shut down until we got five replacements for the ones who went to the hospital. Lying down, every time I closed my eyes, I saw my friend's face peeling off as we removed his mask with the raw, pink skin underneath. The sight of it in my mind made me sick and prevented me from sleeping. My captain and I sat in the kitchen together all night again.

The lieutenant was released from the hospital during the night. He was sent home for a couple of days. The female sub got a cast on her foot and was sent home for a few weeks. The burn patients were kept for IV antibiotics and wound care. I visited the guys in the hospital a couple of times until they were released. They went home for about a month before they could return to duty.

Our backstep singer who had his face and wrist burned so badly was the last to return to duty. His face healed completely. He had scars on his wrists, but he put his watch over the bad side, and it was barely noticeable. He was rather upset that his whole face hadn't been burned. He had small acne scars across his face. The side that burned healed baby smooth and he believed his whole face would be baby smooth instead of a half-and-half mismatch.

The big lieutenant ribbed me about my picture in the paper, "How is it that we all get hurt except for you but you are the one who gets your picture in the paper? Why do we get the pain and you get the glory?"

It was unjust, like life, and no amount of explaining would convince him that I didn't want it or like it. To this day, he beats me up about that picture which is featured on the front cover of this book.

Right after the backdraft, our new chief asked my captain to explain the injuries and his actions on the fire. It was a meeting held behind closed doors, and I will never know what was said, but I did feel the results. My truck crew and captain were declared incompetent and sent for remedial training. The captain was sent for a medical evaluation of his eyes and was not allowed to return to duty until his vision was improved. I understood why my captain hated this chief. We were involved in a terrible fire phenomenon and being blamed, even punished, for it.

A bunch of subs replaced all of the injured people, plus my captain, who was scheduled for eye surgery. In the interim, my truck was commanded by a lieutenant who was newly promoted. The subs had a different routine, and I was disturbed by the whole incident, feeling depressed and grumpy from all the stress. My backstepper laid off sick.

When it came time for our remedial training day, I was the only regular person available. The truck was marked out of service and the sub lieutenant went with me to the academy to chop, throw ladders and crawl around for search and rescue. I performed terribly since I felt sick and beaten. I completed the training and went back to the firehouse grumpier than ever.

Going to bed early, I slept long but fitfully, experiencing nightmares. The sub came in late, letting the door shut hard and waking me. He got up early and moved in and out of the bedroom multiple times, too many times, waking me each time. I got up. Speaking to the sub in the locker room, I told him he should get all of his stuff out of the bedroom in one trip. The sub lieutenant heard us talking and butted in on the conversation. He told me to lighten up. I reminded him that I was resolving a problem according to the book; handle it between the two people involved, and if a resolution can't be reached, then the problem should be taken to the first person in the chain of command. If we needed the next person-him-we would call. The sub lieutenant didn't like that I shooed him away. He wasn't there much longer, because my captain came back and the lieutenant moved on. He didn't seem to like me very much, but frankly, I didn't care.

Chapter 21

It soon felt like home again, with everybody back and healthy. Permanent damage remained though. My captain began talking about retiring. He said the joy of firefighting was gone for him. As soon as he turned 65, and Medicare kicked in, he was walking out of the firehouse for the last time. Since 70 is mandatory retirement age, conceivably he would have five good years left. I loved my captain and our firehouse and attempted to convince him to stay. He was still strong as an ox, very brave, and knowledgeable about firefighting. The surgery to improve his eyesight corrected his one weakness. The new chief, ever harassing us, instituted apparatus inspections and surprise visits like he was going to catch us doing something wrong. We were such a tame house compared to some of the boozing, womanizing houses in the city but he had his eye on us. It seemed we were being singled out because he was a racist, sexist man. I had heard it said about him many times that he doesn't like African-Americans. I have found most bigots are also misogynists. His actions in later years proved me right.

The first fire we had after the backdraft was in a huge, apartment complex. We arrived later than most companies since we were called in extra and it was in another chief's district. Ladders were already being put into the windows, where people were hanging out, crying, and screaming for help. Several people were on ledges, ready to jump, the smoke billowing above their heads through the open windows. A wind shift could send the smoke back into their faces, and they would leap for sure, as they choked on the hot, acrid, foul air.

The chief, to whom we reported, needed the second floor searched for any trapped victims. We suited up, donning our air masks, as my stomach did flip-flops at the thought of entering the burning building. The raw, burned, backstepper's face flashed through my mind. Unable to remove that image planted firmly in my mind, I followed the captain and my crazy backstepper into the smoke. The building was dark, because there was no power. Our flashlights were ineffective against the smoke, like high beams bouncing back in the fog when one is driving. We walked slowly, but still bumped into a lot of stuff. If my co-workers were feeling any fear, I'll never know, because nobody talks much about those things. I kept my mouth shut and concentrated on trying to see through the smoke.

We finished the search on our floor finding no victims. My air was almost gone and so were my partners' air. Normally, I have more air left than the others but with my heart racing from fear, I sucked down the air faster than I normally would.

We found our way back to the entry point, exited, got fresh tanks and reported to the chief for our next assignment. We heard "all clears" being reported from every floor. The fire was under control and out. All of the victims had been rescued. Things calmed down.

The chief told us we could dump our SCBA's and heavy coats. We picked up some of the multitudes of hose lines laid out everywhere. It seemed we loaded a ton of hose. Since we were from another district, we were among the first to be released back to duty. On the ride back, I felt happy that I was able to overcome my fear. I knew I would regain my full confidence with a little time and a few more fires. We laughed with the engine company, back at the firehouse, as we told about our trip and fall and bump into each other exploits in the blinding smoke.

The next duty day, we had an early morning "surprise visit" from the new chief. It wasn't a total surprise to us, because the firehouse where he was based called to warn us as soon as he left. Calling our crew together, he said he had received a bad report about our work on the fire the day before from the chief in charge. He thought we needed more remedial training. After he left, I called the chief who supposedly complained. I was suspicious because the other chief hadn't said a thing to us on the fire and he was the type to let firefighters know right away if he was unhappy with them. He told me, our chief had called him asking about our work on the fire. He said he didn't complain about us in particular, but that he told my chief he was unhappy with everyone's performance, especially the companies that laddered the building. The apartment hadn't been laddered fast enough in his opinion. That was his biggest complaint. We weren't given that assignment. Our job was to search.

My captain went to headquarters and signed a notice of his intent to retire. Being his engineer and in charge of the party for him, I collected donations from everyone. Divorced, I had no use for the quarter carat diamond in my engagement ring. I took my old ring to a jeweler and had the diamond from it placed in a fire department ring made especially for my captain as a parting gift.

On his last day, we had cake and a celebration. His family came, cut the cake for him and took pictures of everybody. He opened the jewelry box and was surprised to find a ring which fit him perfectly. His wife had secretly given us his ring size. In addition, we bought him 20 lottery tickets. He didn't hit the jackpot, but he is a winner in my eyes.

Years later, his family invited me to his surprise 70th birthday party where hundreds of people were in attendance. I was so proud of my captain.

I sometimes see him at union functions and funerals and give him a big hug. My heart swells with gladness when I see him and my other workmates from that station because I can't help but think of the wonderful days we shared. The big lieutenant isn't so big anymore. He lost weight, recently retired and is my publicist.

Chapter 22

Some of the new women coming on and we were told, some of the men, didn't like sharing locker rooms and bedrooms. The rumor was going around that a decision had been made to have separate facilities. A new station was being built and the architectural plans included a women's bunk room and locker room. Nobody asked my opinion. I didn't like this idea. I called the other 13 women on the job, and a lot of them hadn't heard what was happening. We agreed to meet at one of our homes, which was centrally located.

We discussed the situation, getting everyone's input, and arrived at a consensus. The subject of uniforms was brought up. Some of the women wanted to be allowed to wear skirts with our dress uniforms. I could have cared less about that issue, but it was important to some, so I voted to ask the chief to allow it.

We decided to approach the chief and got our calendars out, picking a day when most of us could be there. The ones on duty said they would get somebody to take their place at the firehouse for the few hours we would meet with the chief. We agreed we would ask the chief for separate locker rooms, but keep the sleeping quarters joint. We had many reasons for this. The men were grumbling that we would have private sleeping facilities and be spoiled, while they still had to sleep in big rooms together, with each other's snoring and farting. "There go the women getting special treatment again!" The recent double murder of an ambulance crew of women who had separate sleeping quarters at a firehouse in another county was brought up. We felt that if we stayed quartered with the men, there would be safety in numbers. It would also be less likely and harder for someone to crawl into bed with us, invited or uninvited. Most gripes on both sides were arising from shared locker rooms. People walking in on each other accidentally (we hoped), got the most complaints.

The new chief of the department was African-American. He suffered through the earlier, accepted days of a segregated fire department when African-Americans were confined to one station and a new African-American couldn't be employed until one of those few positions opened. The women expressed how we wanted to be able to talk to the guys almost anywhere in the firehouse. We didn't want them to have too many places to hide from us. After convincing the chief that sexual segregation would be as bad for the fire department as racial segregation was, he took action. Blueprints for the new firehouse were changed to reflect the women's wishes, separate locker room facilities, but no separate bedrooms. We got our skirts too.

Builders were sent to all the stations on the house captains' shifts to choose locations for the women's locker rooms and to decide what modifications were needed. I traded time on the house captain's shift when he called everyone to the kitchen to discuss the remodeling. Even though I had gone along with the majority of the women, I was not happy about it happening at my house. I liked sharing the locker room facilities with the men. After I fired off a smart-aleck remark and was generally crabby at the meeting, the captain cleared the room. Yelling at me, he wanted to know what was my problem. I shot right back at him that I was the one who would have to live with this decision, in isolation from everyone else. I hated to change and didn't want the men to resent my presence by taking up precious space in the house. It was a small place for eight people on a shift and to take a chunk of it for women's quarters would have been a hardship. After our discussion, during which I cried, the house captain refused to let anyone come in and measure for changes. Thanks to him, my station never implemented the required changes.

There were so few women on the job that I rarely got to work with any of them. We were and still are isolated from one another. It was a treat when any of them came through. Ruth was one of the first of the women subs to come to my station. She loved to play tennis and basketball, so we clicked right away. She was the opposite of me on women's quarters, wanting everything separate, but we set that issue aside. She was there long enough that we were able to cement our friendship and have been friends ever since.

She calls me whenever she has a problem wording her treatises. Together, we come up with some pretty, fancy sentences. She compliments me, telling me I am a good writer and encouraged me in this endeavor, writing a book. Ruth had a lot of women friends come to the firehouse and I figured out quickly that she is gay. When I was younger and confronted with it in the army, I didn't understand homosexuality and was afraid of it. Now that I am mature and have been exposed to it, approximately half of the women on the fire department are gay, it is no big deal. I wonder what all of the fuss is about.

Ruth wanted to start a fire department women's softball team. Since, in the past, I had attended many union meetings, and Ruth had not, she asked me to attend a meeting to request money to pay the entrance fee. Standing up in the meeting, I asked the union for money, but the motion was voted down on the floor. We were dismayed because financing was provided for a lot of teams. The men thought it was another special consideration for the women. They said if we wanted to play, we could play on one of their teams.

After the meeting, the union vice president pulled me aside and said I should have never brought that up for a vote. In the future, he told me I should come to the board with a request like that, and they would give us the money. They said they could reimburse us from a special fund that didn't need approval from the general membership. We got our team started and had it for quite a few years, all with money from the union under the table. We had a lot of fun, but it was tough filling a team with so few women available and all working different shifts. We had to forfeit some games. We finally dropped out of the league due to our hectic schedules. Most of us either had day off jobs or kids at home. The newer women haven't expressed interest in a department team.

My son doesn't like Ruth very much, because when he was little, he was pestering her at one of our fire department women's softball games. To teach him a lesson, she stuck him in one of those barrel-like trash cans with bees buzzing around in it. Hearing his cries, I rescued him before he got stung. I was mad at her for a long time about that. She's not very good with children.

My new boss arrived.

A few people called to offer their condolences because, they said, "He is a big asshole."

He had already screwed over several people in order to make captain. One of them told me that when he was going for lieutenant, the evaluation the captain gave him almost killed his chances. The alarm bells in my head rang because I was going for lieutenant. I hoped he wouldn't do that to me.

On the home front, my son was having problems in school. On the advice of a psychiatrist who said he was very intelligent, and probably bored in regular classes, I enrolled him in a private school for advanced kids. It was a small academy, not far from my home, but there was no bus and I couldn't drop him off until 0730. School started at 0800. I lived south and worked north. Most mornings, caught in rush hour traffic, I barely made it on time. We had to be at work by 0755. One of the guys on the C shift agreed to stay over for me whenever I was late. The captain didn't like it and although he couldn't do anything about it since I had my shift covered, he raked me over the coals anytime it happened. Very seldom was I late and then only five or ten minutes. To this day, I am thankful that the C shifter helped me out with my kid.

Truck work is heavy duty stuff. The captain seemed to have his doubts that I could be a good truckie. The first month he was there he had all of us on the truck haul the ladders off, including the 35-foot beast. He ordered me and my crazy backstepper buddy, who is about the same size as me, to raise the big ladder by ourselves. This being my third year on a truck, I had raised a lot of ladders. With no problems, I put that heavy thing up, and brought her back down lickety-split. The captain, a martial arts instructor, considered himself in excellent shape, and a good firefighter, but he came from an engine and hadn't raised a ladder in a while. He placed himself at the heavy end and had the sub foot it. Since his technique was rusty, and he was out of practice, he struggled in the middle at its heaviest point but managed to force it up with brute strength.

After he wiped the sweat off his face from exerting himself, he turned to me and said, "Kathy, you are one strong bitch."

Nobody laughed. We all stared at him.

He stammered, "I mean that in a good way. That ladder is really heavy, and you made it look easy. I have to respect that."

We had a few little fires together. We got along okay and I thought we worked well together. Evaluation time arrived. Talking to other firefighters in the promotion process, they told me they were getting great reviews with their scores all nines and tens. Evaluations were set up on a scale of zero to ten. A zero meant you were dead, a five was average and ten meant you wrote the book on firefighting. The officers of the firefighters going for rank were giving evaluations that made the candidates look like saints, earning them the most possible points toward promotion. They wrote the necessary glory stories to justify the high scores.

Just like he did to his other subordinate, the captain gave me all fives and sixes. Arguing with him that it would hurt my chances in the promotion process, I urged him to change my scores. He said, as a new captain, it could cause him grief if he were to write an evaluation with high scores. In a meeting he attended, the officers were told that fives and sixes were the standards. Only an exceptional person should receive a better score. He assured me that the scores would all be lower for everyone on the evaluations this time. Having spoken with many people, I knew this wasn't true. My captain and I went round and round, but he refused to budge. I refused to sign the evaluation. He sent it in with a note explaining my refusal to sign. It was entered into my file.

The promotion process is almost as long as the hiring process and has almost as many facets. Historically, each process has changed to suit the administration's whims. In this process, every firehouse was provided several books and other study materials from which the test questions would be derived. We would be tested on the general orders too, and since I had learned them in the fire academy, that part was review for me. Studying the books while on duty, I highlighted everything of interest, put comments in the borders, and generally marked them up. Since I was the only one going for promotion at my firehouse, there weren't any complaints.

After months of preparation: studying, personnel file updating, testing, and interviews; my scores came back. I kicked butt. My overall placement on the promotion list was number seven out of more than 100, and I was number one in several of the categories. My evaluation, as I had feared, dragged me down.

My excitement rose as I discovered a significant mistake in my score. As an engineer I should have received 50 points in a category where there was room for improvement. If it did not make me number one, it would move me up a few places for sure. The scores were derived through a complex formula and 50 points in the beginning translated to maybe a point or less at the end. The top 10 people were separated by fractions of a point. I couldn't wait to see how this would affect my standing.

Bounding over and calling on the Centrex, I euphorically told a person at headquarters about the oversight. I was transferred to a chief and informed that I would have to appeal any mistakes in scoring by expressing my complaint in writing at a special appeal process. Missing only seven test questions, I planned to appeal six of them because I found support for my answers in the books. Although I had the top score on the written test, I could still improve my standing with a better grade on it.

Appearing at the appeal session armed with notes and my engineer certification, I stayed the full three hours. A chief reviewed my scores with me, showing me where I could improve. He congratulated me on my shining performance, but was concerned about my evaluation scores. He pointed out the dichotomy of my stellar performance in every facet of the procedure with the exception of the evaluation scores, where I had placed almost last. I told him who my officer was and he shook his head in understanding, not making a comment. Carefully I documented page numbers and quotes from the books supporting my appeals. By the time I finished, the appeals proctor, a civilian hired by the department to oversee the process, and I were the only two people left in the room. As he packed the books in a box, I handed him the appeals which he threw in the box on top of the books.

I anxiously awaited my new score. Ripping open the envelope, I was disillusioned to see it had remained the same. Reviewing it, I discovered I still hadn't received my engineer certification points. When I called headquarters and talked to a chief, he said the score was correct. I told him I did not receive the points I deserved. He said I should have appealed it. I swore that I definitely had but he denied receiving any appeals from me. I told him I had witnesses who could swear they saw me at the appeal process. He said the department wasn't denying I had been there. They were denying I made any appeals. I asked him why anyone would attend the appeal process if they weren't going to appeal anything.

He said, "There were firefighters who attended to review their scores and didn't appeal anything and you were one of them."

I described each appeal, remembering vividly what I had written about every test question I felt was graded wrong. They refused to listen to me. I went to the union who said they couldn't help me because I had no proof. I contacted a lawyer referred to me by a woman on the department who, she said, helped her win a discrimination suit against the administration. Her lawyer said she couldn't help me since I had no proof.

I was despondent and livid that the administration would stoop to cheating to prevent me from being promoted. I consoled myself with the fact that I ranked high enough on the list, it would only be a matter of time before they would have to make me a lieutenant. Fed up with my new captain and his selfish ways, after all, he helped them keep me down with the evaluation he had given me, I decided to put in a move. It was atypical behavior on the fire department to not help your own people. I didn't want to work for a man who cared little for his subordinates. His actions wouldn't have mattered that much if someone in the administration hadn't ripped me off for my engineer points. Since the whole situation was a bitter pill for me to swallow, I couldn't stand to be in the captain's company anymore. Requesting a south side company where I would be closer to home and have a shorter drive to work after dropping off my son, I was awarded the position based on seniority.

Upon receiving my moving orders, I got the map out again to study my new district. After a few work days, I packed all of my fire department belongings, including my "I like Dick" pin that an anonymous someone had pinned to the towel on my locker. It was a button from Richard Nixon's campaign for president. I don't think it was intended as a confirmation of Nixon, though. I considered it humorous and saved it for years.

Part Four

A New Home

Chapter 23

My new firehouse was in a poor, white neighborhood. A lot of inadequately educated, hard-drinking, people resided in the area. I wondered if there was inbreeding or poor medical care, because I never saw so many cross-eyed, deformed-limbed people in all my life. My new lieutenant was the one with whom I had argued when he butted into my conversation with the sub at my old station. We weren't starting out on the best of terms. The engine had all new people except one firefighter who had been an academy classmate of mine. He was on sick leave because his girlfriend's, ex-boyfriend had beaten him over the head with a log from her fireplace. The injury was so severe, it required the insertion of a metal plate in his skull. He wouldn't be back to work any time soon. Happily, Ruth was at this station on the B shift.

Studying the district on my map every day wasn't helping much. As I said before, looking on a map and getting there in an emergency vehicle are totally different things. My old firehouse was on the south side of the street, so I turned right to go east. Here, I turned left to go east. The first few months, I instinctively turned right when an east address came out. North and south were opposites from my old station too. For some inexplicable reason, I confused State and East streets. When I explained to the chief and his aide, who were stationed there, that both streets had the same letters except for one, and my brain scrambled them sometimes, they looked at me like I was crazy. Jokes abounded for quite some time about my dyslexic confusion.

When I turned west out of the firehouse, thinking I was going east, I would have to make a big U-turn around the fountain. I was usually going pretty fast, because the lieutenant was yelling at me to get headed in the right direction. The backsteppers said, I created unbelievable G forces that pulled them from their seats, causing them, they claimed, to go temporarily blind and lose consciousness. G forces eventually became known as Gillette forces.

After four months, the backstepper who had his head bashed in returned to duty. Lonely, because nobody here talked to me and missing my friends from my old station, I looked forward to his return. He was the one who told Mandy he would never say anything bad about me after I had raised the ladder he was supposed to be helping with. The lieutenant hadn't forgiven our first encounter and was giving me the cold shoulder. The new backstepper took his lead. The chief, with his aide, stayed in their quarters out of the controversy.

The meals cooked by my co-workers were fatty and unhealthy. I went to the doctor and obtained a physician's written order, excusing me from firehouse meals, the only way stipulated in the general orders to be exempt from meals. I didn't want to gain weight, and that would never happen eating their cooking. Making myself tuna fish sandwiches and munching on apples, I would sit with them at meal time. Afterward, I helped clean the kitchen, but the lieutenant hated it. He said a house that eats together, stays together.

When my classmate was released back to duty, he was nice to me for about an hour and then took the lieutenant's side. He actually began badmouthing me worse than anyone. After about six months, I got tired of their crap and told them, in true firefighter fashion, they could shove their nasty comments up their asses.

Every morning, before I washed the engine, I put everybody's boots and coats up so they wouldn't get wet. After I told them off, I said they could put their own damn gear up. When my classmate forgot, I sprayed his stuff right along with the engine. He busted a gusset when he saw his wet gear, and ran with his dripping coat to show the lieutenant what a bad person I was. The lieutenant nearly fell over laughing. I could hear him in the office guffawing from where I was standing in the apparatus bay. His laughter echoed through the entire house.

After that incident, my officer figured he couldn't run me off and I wasn't a pushover. He began to include me in everything. This made the other backstepper grudgingly accept me, but my classmate never stopped complaining. The lieutenant went to a nutritionist, obtained a healthy-cooking booklet, and mandated wholesome food. The chief chimed in that it was a good idea because his doctor told him he needed to lower his cholesterol. Things finally started to turn around with my classmate being the only holdout. Wetting his gear down regularly, the grudge-water-fights turned into a contest between the two of us. I would wet his gear down, then he would wet me down. Eventually, the rest of the engine crew joined in, and they all ganged up on me.

One morning when I didn't feel like getting wet, I ran behind the chief who, minding his own business, was reading the morning paper. They dumped a bucket of water over my head, getting the chief and his paper all wet. They were squirting me with their super water soakers as I ran up the steps. They were still laughing when I returned until they saw the hose in my hand. I had fired up the engine and pulled the red line (a smaller diameter hose used mainly for car and trash fires that was wound on a reel and easy for one person to handle even at 100 pounds of pressure). Frozen in position, rooted to their spots, their mouths hung open. Never, did they believe, would I open the nozzle in the kitchen.

I shouted, "Peace by superior fire power!", and let them have it as I advanced the line into the kitchen.

After thoroughly soaking the lieutenant and the other backstepper, I turned and followed my classmate up the stairs, bounding up them two at a time like a lion after her prey. He had snuck around behind me but I was going to get him. I saw him dive under the chief's buggy (a small, van-like vehicle) to hide. Opening the line again, I shot the water under the vehicle. He tried in vain to get out the opposite side, bumping his metal head a couple of times. After I allowed him to escape, he joined the others in the only safe spot in the house, the men's locker room. (This house had separate locker rooms.) After putting the hose away, I shut down the engine. The place was pretty well stunk-up from the diesel fumes. Even though I probably lost a few lung cells, I was happy. I paid them back double, triple, no, quadruple. Nobody wet me down after that because it was a pain in the ass for everybody to clean up all that water in the kitchen. Nobody wanted a repeat of that debacle. The chief loved the way I fought back and tells that story frequently.

We soon moved from water fights to ping-pong tournaments. It was the front, the lieutenant and me, against the back, the two backsteppers. We kicked their butts so many times (they never won a game), that the one finally refused to play. After throwing the paddle in frustration, the one couldn't be dragged back to the table.

For dishes, we drew cards or rolled dice deciding by luck who would wash and dry. I was a fortunate, son of a gun, and hardly ever had to wash or dry. Skipping off to rub it in, I would get the mop and bucket to clean the floor. Even though I won fair and square, it pissed them off royally and they called for my head.

The chief's aide was puzzled by their griping. He questioned them frequently, "How can you get mad at her, when you invented the game, made the rules, and she beats you at it?"

This was a busy firehouse averaging eight to nine runs a day. I would guess that 50 percent of the emergency runs involved alcohol. Drinking seemed to go on into the wee hours. Due to that, we never got to sleep through the night. We would usually get three to four runs, sometimes important but sometimes goofy, after midnight.

The lieutenant, opting for the opposite lifestyle, did not tolerate wine, women, gambling, or getting away. We were a straight-up firehouse in an ocean of carousing, angry, put-up-your-dukes-and-fight people. We went on numerous stabbings and a few shootings where our patients refused police help saying they would take care of the assailant themselves, extracting their own revenge. Teenage pregnancy was rampant. The number of 12 and 13 year old girls with babies in their bellies was astonishing. Many of them appeared to be living out of cars. We would often arrive at a street corner to find a young girl with a much, older man in his car. She would be frightened, and in pain, pleading for us to get that thing (baby) out of her. Sometimes, the man was an uncle-daddy. I thought of the poor baby, destined to struggle through life, raised by a young, lonely, insecure, penniless girl; kids raising kids.

Tragic fires happened too frequently in this neighborhood. Some were smoking in bed or children playing with matches and then trying to hide from the fire. Some were arson: gang members, jealous boyfriends, vengeful neighbors. Some were accidental: candles, incense, unattended food on the stove. Several times we found people who tried to put out the fires themselves. They were burnt to a crisp lying on top of their fire extinguishers.

Chapter 24

Around 0200, early one morning, my company was dispatched to a fire in a government subsidized, apartment complex. We were the third engine due and were almost there, when the first arriving company reported a fully involved, two story, brick, multi-unit building. They started hooking up as the second arriving engine pulled around the back in the grass. We stopped at the curb, and everybody jumped out. My classmate excitedly pulled a preconnected line and started to run off without unhooking it. Yelling at him to stop before he lost the whole lay off his shoulder; I unscrewed it from the engine.

We all hurried off to the fire. I went to help the other engineer in the back get his engine operating, which was my duty. The rest of my crew went up a ladder to the second story window. They attempted to enter where flames, shooting 10 feet out the window, hampered their entrance. Finally getting some water through the window onto the roaring fire, they painstakingly beat it back and climbed through.

Two people had jumped out of the window before our arrival. They were found on the ground, both unable to move. The ambulance crew needed help getting the patients onto back boards and into the ambulance and drafted me. The parents were concerned about their daughter, who wouldn't jump and was still up there. The dad, his leg mangled, said he wrapped his son in a sheet and jumped out the window with him, but we couldn't find the little boy outside anywhere. The mom, suffering a broken back, begged us, as she was wheeled away on the cot, to help her children.

When I came back from loading the dad, my crew came down the ladder with a teenage girl. She was not breathing and had no pulse. My classmate laid her on the ground and went back up the ladder. Mary was on the squad with the second arriving company. She and I started CPR on the teenage girl. We ripped her shirt off and Mary started compressions on her chest. I got out the bag valve mask and breathed for her. Badly burned, the poor girl's skin started coming off her chest getting all over Mary's gloves. She was having difficulty keeping her hands in the correct CPR position. She said the girl's skin was not normal at all and felt very slippery.

Our patient was an African-American with beautiful, chocolate colored, skin, but wherever Mary touched, the girl turned white because her skin was sloughing off. Mary's gloves were now covered with minute balls of skin. It was horrifying to watch this happen, but wanting desperately to save her life, we kept working her. A paramedic arrived, and put a breathing tube down her throat to protect her airway and deliver better oxygen.

Just then, my backstepper brought out her brother, a little, four-year-old. He must have fallen out of the sheet as his father frantically jumped. They found him directly under the window. As my backstepper carried him under his arm, the boy's lifeless legs, arms, and head were dangling, swinging like a rag doll's, as the firefighter hurriedly walked toward us. He was badly burned and the other paramedic pronounced him dead immediately with no resuscitation efforts attempted.

Our paramedic quickly reverted his attention to our patient, administering her heart starting medicines as we put her on a gurney to take her to the ambulance. Mary and I boarded the ambulance to assist. On the way to the hospital, after several rounds of medicine, her heartbeat returned. We were elated as the monitor showed a rhythm and we felt a pulse. With the image of her baby brother vivid in my mind, I was grateful that she was just now returning to consciousness and did not have to see her burnt sibling. Focusing concentration on her, I pushed her little brother's image out of my mind. We transported her to Wishard, which is much further than the closest hospital, but with its renowned burn unit, we knew she would get the best care there. By the time we arrived, she was trying to move her hands. It seemed like she wanted to ask us questions.

An ambulance returned Mary and me to the fire scene. Things had settled down and people were quietly picking up hose. No one likes to see little kids burned up, so the usual banter was missing. After picking up the hose, I was leaning against the building when one of the men came over next to me and lit up a cigarette, offering me one. I hadn't had one for more than three years but I took it. I needed something. I really needed one.

My lieutenant walked over to me staring at the cigarette and said, "I didn't know you smoked!"

"I don't, but after tonight, I can't live without one!"

He asked me about the girl. Relating the events in the back of the ambulance, I was pleased to report that she was alive.

It was 0630 before we got back to the firehouse. The guys all went to their locker room for a shower. I went to the kitchen to get some much needed coffee. The television was on. The news showed the fire we had just put out. They reported two critically wounded and two dead. What? Two dead? They got it wrong as usual. I called to set them straight. Correcting them, I said it was true we had initially found her dead but brought her back. They got it mixed up. They didn't know about us saving her.

They said, "We're sorry. She died later at the hospital."

I slammed down the phone. I was sick and tired of reporters getting the story messed up. To dispel my lingering doubts, I called the ambulance crew. They hesitated, then confirmed she had died. She woke up and asked about her mom, dad and brother. They told her she could see her parents soon and they were in the hospital waiting to see her. She passed away shortly after that.

Hanging up the phone, I ran upstairs to my locker room. Shutting the door and locking it, I slid down the wall, crying, muffling my sobs. No way did I want the guys to hear me weeping. I snuck out the back as soon as my relief arrived. My normal routine was to give a report to the oncoming engineer, but I couldn't give a report with my eyes red and swollen. I just wanted to get home. I was living with a guy who had two kids. After getting my son and our other two off to school, I lay down to sleep. Exhausted, but unable to rest, I got up and cleaned house. Everybody got extra hugs when they came home from school. They looked at me strangely, wondering what was wrong with me. I couldn't tell them why I was so clingy. It was too gruesome for little ears.

At work, seeking solace by doing the usual routine, I put my gear on the engine. My lieutenant, sensing my reticence, gently asked me if I needed some counseling. Not wanting to admit any weakness, I told him no, definitely not. We got an early medical run. The lieutenant's coat was thrown in his seat. I put it there as usual so I could wash the engine. Ready to go, I was sitting in my seat. He climbed up the step, grabbing his coat to move it out of the way.

He stopped and looked at his coat asking, "What is all this stuff on the side of my coat?"

I looked down to see what he was talking about and realized it was skin, the burnt skin, which rubbed off the girl's body when he carried her over to us. Having seen it on Mary's gloves, I recognized it immediately. I gasped and looked into his eyes. He understood from my reaction, what was on his coat and now his hands. He started saying, uhhg, uhhg, uhhg, and kept repeating it as he dropped the coat on the floor. He ran to the bathroom, washed his hands, grabbed another coat off the rack and got on. He was pale and stared straight ahead. I asked him if he was okay. He nodded yes.

When we returned, he got a brush, scrubbing and hosing down his coat. He scrubbed that thing for a long time, flipping it over and over to make sure it was clean. He went into the office and didn't come out until lunchtime. He didn't eat much, and later said he had a bad case of heartburn. I gave him antacid from our refrigerator stock. He barely spoke the rest of the day.

The two backsteppers wrote an official report about the lieutenant's bravery and how the three of them found the two victims. At our annual awards ceremony, the lieutenant was honored with Firefighter of the Year. My classmate was awarded Rescue of the Year. They both received money as part of the awards. The lieutenant used the money to take our crew and significant others to an expensive restaurant for dinner.

The next day at work, he asked me how I liked the meal. I told him I had really enjoyed it and that he was very nice to treat everybody with his hard earned award money. He certainly didn't have to do that. He said he couldn't have rescued those kids without us, therefore, he felt we deserved a part of the award. He told me he was feeling a little guilty because of the way things happened. It wasn't his idea to submit the official report, and he hadn't written it. (I knew that because I had signed it after our backstepper wrote it.) He said he was unhappy with it because it made my classmate seem fearless.

The real story was; he had to practically push my classmate into the window. My classmate climbed to the top of the ladder and hesitated. The lieutenant said he had to get his shoulder under my classmate's butt and raise him up by using the strength in his legs. My classmate tumbled into the window landing next to the girl. The other two climbed in the room, opened the nozzle and put out more fire. My classmate and the lieutenant climbed down the ladder with the girl, while the other backstepper stayed in the burning room, putting out more fire. The lieutenant hurried back to make sure our other backstepper was okay. Finding the boy, the backstepper carried the rag-doll-like kid down the ladder.

The lieutenant said, "If you listen to his version, he feared nothing and was the big hero. I shouldn't have let him get that award, and you should have received more recognition. You are part of the team. If it hadn't been for you, stopping him and unscrewing the connection, he would have lost his shoulder lay. He is too excitable."

I was surprised to hear all of that. It seemed my lieutenant had finally gotten over our first fight when we were both at the north side company. In defense of my classmate, he was not the one who wrote the report. The other backstepper wrote it, but he didn't correct it or dispute it and he bragged after he received the award. I think the thing that truly bothered my lieutenant, was that my classmate had still not incorporated me into his concept of the team. My lieutenant wanted that most for us, to be a team.

Chapter 25

Sometimes we shook our heads in disbelief at how our neighbors lived. They had poor ways. We would be called out at midnight, two or three in the morning. As we rode down the street, people and their kids were outside on their porches. A lot of the houses didn't have air conditioning so it was too hot to sleep. It made you feel sorry for them until you entered one of these houses which were smelly, filthy, and buggy. They were buggy because rotting food was left out on counters; dirty dishes were left on the table and in the sink; garbage was everywhere. Chunks of animal feces from innumerable pets were slovenly left, soiling the carpets. The stench could be unbearable. Roaches were crawling around. Sometimes they would crawl on the people. The people were so used to having them around they would just flick them off.

We were grossed out. We left the houses saying critter check, stomping our feet before boarding the engine. We didn't want to bring any of their bugs back to the firehouse. Cockroaches contribute to asthma. No wonder so many of these kids have asthma problems.

Too often, we went to houses after midnight, where the parents were drinking, partying, and smoking. Young kids were asleep on the floor, covered with blankets, in front of the TV turned up full blast. School was bright and early. I wondered how these kids could stay awake for their teachers, especially when they were awakened in the middle of the night with a rough kick and a loud order from Dad to move out of the way so we could come through with our medical equipment. The ambulance would arrive next with a cot so the kids would have to be awakened and moved again.

The streets in this area are very narrow. Few houses have garages. Cars line the streets, parked along the curb.

Sometimes both sides of the street have parked cars leaving only a narrow middle lane through which the big apparatus has to pass. We were dispatched on a run, and I had to travel down one of the narrowest streets. A truck with bowing, wooden-slat, sidewalls built into the bed was parked on the right side of the street. A row of cars was parked on the left side. As I tried to pass, our mirror, which sticks out considerably, smacked the truck slats and broke upon impact. I hit the brakes and stopped instantly. The lieutenant got on the radio and reported the accident. Another engine was started for the run. The police and a different chief came to take pictures because our regular chief was out on a fire.

My classmate had a field day, chiding me for my bad driving. He insulted me in front of curious civilians who came out of their houses to see what was happening. When the alternate chief got there, he pulled out a tape measure because he saw I had only inches on each side. After he measured the distance between the vehicles through which I passed and determined the space was too small, he conferred with the police and the officer issued a ticket to the owner of the truck because his sideboards created a hazard. My classmate abated. I felt sorry for the owners of the truck.

Back at the station, I was sitting at a table by myself in a gloom and doom mood. My lieutenant passed by, stopped, and came back.

He said, "Don't be so hard on yourself. It was just a mirror and everybody has accidents. Don't let your classmate get to you."

The shops came to put a used mirror on at the station. We were out of service approximately 15 minutes while they bolted the mirror through the door. It was makeshift, but it worked.

For the boo-boo to the mirror, I had to face another accident review board which heard all cases involving damages over $250. The lieutenant and I waited in the hall at the mother house for an hour. When it was our turn, the alternate chief and my lieutenant appeared as witnesses. They argued that for a broken bracket and a used mirror which the shops put back on with two screws, the damages should be less than $250. With such formidable opposition, my lieutenant and a battalion chief, the board agreed to lower the dollar amount and dismissed the case. We were in there five minutes. Thanks once again for your support, gentlemen.

Since the lieutenant and I couldn't find anyone who would play ping-pong with us, we invited an ambulance crew to stop in for volleyball. We strung the net from fence to firehouse, moved all of the cars out of the parking lot and played three on three. The chief and his aide wouldn't join us. Part of our playing field contained a 10-foot stairwell to which the ball seemed magnetically attracted. We took turns climbing down the steps to recover the ball from there, fishing it out of a permanent puddle of water, wiping it dry on our uniforms. A few times we knocked it clear out of the parking lot and would have to go out of the gate to retrieve it from the parking lot next door.

We played jungle ball just like my subbing days, no setting and no spiking, just get the ball over the net. Like most firehouse courts, it was asphalt. There was definitely no fancy volleyball moves like digging for the ball unless you wanted an inch of gravel buried in your elbows or knees. Our games were interrupted a lot by our or the ambulance's sudden departure for runs.

We left our gate unlocked and open so we could come and go easily. We got a rash of false alarms. When we came back, the guys said they had been robbed. Wallets were missing, and so was the locker room radio. I ran to check for my purse in the women's locker room. Luckily, it wasn't touched. The thieves probably weren't aware that my locker room was right across from the men's. I always kept the door that leads outside locked just in case somebody wanted to "accidentally" barge in attempting to pay me back for some prank. Ruth found pictures of nude firefighters shoved under our door and gave them to me saying I would appreciate them more than she. We never complained to the administration about that, but it made us a little wary, so we kept the doors battened down.

We started locking the gate and all the doors attempting to prevent more break-ins. Several times at night I couldn't get out of the gate to take out the garbage because I forgot the combination. One night, I heaved the garbage over the fence, and it caught on the barbed wire top, ripping a hole in the sack. I had to go back in, get the combination, then clean up the mess. A few months later we became lax about locking the gate until a brazen thief came in the front door during lunch, saw no one was around, and walked off with the VCR. We locked up again.

One of our runs was to a house with a porch that had three steps leading up to the front door. When we arrived, we found a woman at the bottom of them with her daughter waiting for us. The woman, our patient, was barefoot, sitting and holding her leg gingerly. She had a nasty fracture with a floppy foot. She was in a lot of pain, but calm. We got our scissors so we could cut her pants and apply a splint. She started getting upset because she had just bought the pants and didn't want them ruined. She said it had been years since she was able to buy anything new and could we, please, save her pants. The ambulance crew got there, and she was now crying. They reasoned with her that the pants had to be cut off to apply a splint and ease the pain, but she cried harder. We left the pants alone, applying the splint over them and let the hospital deal with it. I thought it was pretty sad. She was more concerned about her clothes than her messed up ankle. As a mom, I can understand. When you don't have much money, and have kids, Mom is inevitably the last one to get anything new.

We were dispatched to another porch house. Two elderly people, a man and a woman, were on the front porch screaming and yelling at each other. Her hair was totally white and looked like demon hair, sticking out every which way from being yanked. He was bald. She had a frying pan in her hand, swinging at him. It was one of those black, iron skillets that weigh a ton. She landed a hit before we arrived which swelled into a humongous goose egg on his shiny head. He had a handful of white hair that he snatched off her head. He was shaking it at her, yelling that he was going to rip the rest of it out if she didn't put down the skillet.

Our arrival on the scene didn't stop a thing. The police had to wrestle both of them. They claimed they had been fighting for 50 years. I believed it because they dodged the police lithely despite their advanced years. The officers called for backup in order to get them under control. They blamed each other for getting handcuffed and spit at each other across the porch. Everybody was dodging the spit balls. We separated them by walking him across the street. Out of eyesight and spitting distance, they finally calmed down.

One night, around 0200, we were dispatched to a person with chest pain. We got at least one run a week where somebody was faking seizures, passing out, or something similar. When awakened for stupid, fake runs, we get irritated. My classmate never hid his disgust from the fakers. We would knock on the door of the house and be led to the patient. One time, a man was sitting on the edge of his bed, clutching his chest, claiming he was in terrible pain. He gasped and shot straight backwards, flopping on the bed.

My classmate groaned and said, "I am not impressed with your theatrics."

He gave him a sternal rub (a knuckle rub on the bone above the heart, which is very irritating and impossible to ignore) and told him to sit up and talk to us. We were staring and waiting. I thought he was faking too. The way he flopped back seemed unrealistic.

The lieutenant checked for a pulse at the carotid artery in the neck and said, "He's not faking. Let's get him on the floor and start CPR."

We made every effort to bring him back but he didn't make it. My classmate didn't hear the end of that. Every patient we had, he was told to check for a pulse. That little bit of teasing put him over the edge. As the saying goes, he could dish it out but couldn't take it. He went to the chief and lieutenant and told them I had to go or he was going. They told him good-bye, adios, see ya', don't let the door hit you in the ass on your way out.

Two months later he was at a new station where he promptly backed the engine into the firehouse. My lieutenant gleefully reported this little accident to me. We snickered for days over this since he had broadcast my accident all over the battalion. I was beginning to really like my lieutenant.

Anytime a fire occurs, it is a big event, especially in this neighborhood. Being a rarity, a female firefighter, I was always a celebrity at the scene. People asked for my autograph; if I would pose with them for a picture; they wanted to shake my hand; pat my back; give me the thumbs up, and atta girl. With an adoring crowd around me, I waved to my lonely guys standing off to the side by themselves. My fans about keeled over when they saw me climb in the driver's seat and take off, "a little girl like you, driving that big thing!" The guys would get jealous because I got all of the attention. The men could be carrying victims out of a fire, and they would only get a glance. Kids would come visit me at the firehouse almost every day. I gave them tours and showed them the firefighter carry by throwing them over my shoulders.

Two young boys, one of whom had a major crush on me, brought candy and presents when they visited. He brought me a ring from a Cracker Jack box and proposed. He wanted to marry me when he grew up. Those boys rode their bicycles to the firehouse in rain, sleet, and snow. One day they were chased by a car. They were afraid the stranger was trying to abduct them. The lieutenant and I called the police. The boys gave a good description of the car and the man inside. After the police left, my lieutenant said they couldn't walk or ride their bikes to the station anymore. He required them to have a parent bring them and pick them up. I didn't see them much anymore. I still remember them and wonder how they are doing. I thought they were going to cry as they left the station the last time I saw them.

The lieutenant and I had a big fight over this. I believed it was good for them to have a positive influence in their lives, imagining and striving to be firefighters. Plus, they were honest, telling their fears about being chased. How would that affect them? Would they hesitate to seek help the next time they needed it? The lieutenant was adamant he would not have their blood on his hands. He said they would have to have parental supervision to and from the station or they could not come.

Two years had passed since the first lieutenant was made off the promotion list. Five people had been promoted, and number six retired. I was next. They could delay my promotion with their cheating ways, but time was on my side. The list would be good for another year or two. I waited.

One night we were awakened for a fire in an apartment complex. The chief got out first in his little buggy that can really go. Our turbo-charged engine barely kept up. The first-in company reported smoke showing and a working fire so we caught a plug, laying a second line. Apartment fires can spread and turn nasty quickly. We were prepared. This complex was called Brick City. It was a government-subsidized housing program, which was on the down side of its lifecycle. Many of the apartments were boarded shut, but there were still some residents around. While I was minding my pump panel, another run was dispatched to the downtown area, which are generally false alarms. A lot of companies were on the initial dispatch because this was a high rise alarm. It surprised me when smoke was reported.

Our little incident turned out to be somebody smoking in bed, a mattress fire, quickly extinguished. There was an unusual amount of talk on the radio. It seemed all hell was breaking loose at the downtown fire. I was sure I heard somebody on the radio calling for help. Dispatch was calling a truck's radio to check if their emergency status was real. We have small orange buttons on our radios to press if we get into trouble and need help. They are recessed and hard to push, especially with our bulky gloves, but accidental activation of the emergency status buttons are frequent enough that verification of a true emergency is sought. There was no reply to control that the button push was accidental, so it was assumed to be real. Many more companies were being dispatched, and we would have been one of them if we weren't at this mattress fire.

Hearing the mess downtown, my crew came running out urging me to help get our water shut down so we could pick up the lines and get back in service. It is not something that can be done quickly, no matter how much you hustle. We accomplished it as fast as possible and marked back in service. We were at the far edge of our district, and we would have a long ride to the downtown fire. Nonetheless, we were eager to go. Radio chatter died down quickly, and it was at a minimum now. We stayed up for a short while listening but you can't tell much from our transmissions since most fireground talk is done face-to-face. We went to bed.

Hours later we were awakened by the loud, obnoxious sound of the Centrex, our official phone. The backstepper answered and was informed that two firefighters, a sub and a long time veteran, were dead, along with a civilian; and five firefighters were in the hospital from the third alarm fire downtown. God, I thought I was going to puke right there. I swallowed hard, tasting the bile in the back of my throat. All of us knew both of the fallen firefighters. I imagined my crew members were feeling as sick as I was. We were dispatched to the scene to relieve some of the companies who had been there for hours. They were despondent and needed a break. Our department consists of about 800 people, a relatively small, close knit community. Everybody knows everybody. When tragedy like this hits, it feels terribly personal.

We went to the firegrounds without lights and sirens. Hoses were still laid out, although some had been picked up. My officer told me to stay with the engine, even though I wasn't pumping or hooked up. I did what I was told without protest. Sitting in the engine by myself, I felt nauseated. I swallowed hard to keep from puking. There was no radio traffic, nothing to keep my mind from my stomach. I got out to breathe some fresh air. The cool breeze felt good. I went around the back of the engine, away from the smelly, exhaust pipe. I heard somebody quickly approaching. Looking around the corner, I saw the chief's aide bent over holding his knees. He was drawing deep breaths. I went to him and touched his shoulder. He turned, startled, and looked at me.

He put one arm around my neck and started sobbing, "Oh my God, that poor young firefighter with his wife and little baby at home. Now he's dead. She'll have to raise that baby by herself. His little girl will never get to know her daddy. God, what an unthinkable tragedy!"

He cried a while longer, then blew his nose on his handkerchief, and lit a cigarette offering me one. I constantly bummed cigarettes from him, hating to buy any because then I smoked more. My stomach was too topsy-turvy to smoke. I declined. He walked off. I hurried behind the engine and puked.

It was daylight before my crew came back. They were given the grisly task of moving the dead body of the civilian from the hallway. The body of our young, sub firefighter was left crumpled on the floor where he died, for the purpose of investigation. The chiefs in administration had decided to block everyone from the area. It made some of the guys fighting mad. I heard some threats and shouting occurred, but no one was disciplined due to the emotional trauma everyone was suffering. The veteran firefighter had been taken out on a stretcher early in the incident with CPR being performed. I saw it on the news.

My crew told me the fire room was eerie looking. The ceiling was totally burned out. All of the tabletops were charred, but they were still standing. Fallen debris was everywhere on the floor. It was a big dining room with two false ceilings, where the fire roared through unimpeded and undetected, causing the room to heat up and the gases to flashover. The instantaneous ball of flames caught the firefighters off guard. They had seconds to get out or die. The ones closest to exits with hose lines to follow, survived. The two that couldn't bail fast enough, died. The one, who activated the emergency button and was rescued by the company that arrived after the flash-over, was hospitalized for months, barely clinging to life. The crew that pulled him from the fire said, they needed to wet him down before they could even touch him much less drag him out because whenever they touched him, he cried out in pain. They were causing severe burns by pressing his superheated gear against his skin.

All of us at the fire, no matter how small our involvement, were counseled and debriefed. It was mandatory for everyone to attend at least one session. We were encouraged to ask for more help if needed. The session I attended had the crew who rescued the surviving firefighter. The backstepper, Gary, we were in the academy together, confided that his boss, a lieutenant, bravely entered the flaming room, put out the fire, and found our badly burned survivor.

He said, "The lieutenant is my hero."

I told him that he, having stayed by his lieutenant's side during the entire incident and his whole crew are heroes and the downed firefighter owed them his life. He squirmed at that thought and tried to downplay it.

The funerals for our two, deceased firefighters, killed from smoke inhalation, were attended by thousands. The fire made national news because the Mike Tyson rape trial was being held in Indianapolis with the inevitable media circus. The media reported the jurors lives were saved by our firefighters who led them to safety.

I felt sick. I watched the procession on TV in the family room. The kids, coming home from school, were concerned when they saw me lying on the couch, a position I rarely took. I explained the tragedy as I cried and got comforting hugs. Kids can be so sweet when they are not being little devils.

The union sponsored a wake at a local bar. Unable to sleep, I went late and met firefighters from all over the continent. An officer from Canada and I traded belt buckles. I traded department patches with some other guys. I lost the patches but wear the belt buckle still even though the Canadian emblem fell off years ago. It is gold, which is a chief's color. When anyone notices and points out that I should be wearing a silver buckle, I tell them how I got it. No one has ever made me change it.

A study of the high rise fire was commissioned by the department. My chief, assigned by the department to oversee the process, brought in outside experts who were paid to examine every facet. The administration wanted to take positive steps toward preventing another horrible event like that. The report was hundreds of pages long. The recommendations were implemented to ensure our safety, as well as it can be ensured, considering the nature of our job. High rise fire fighting is very hazardous and tricky. Many lives are lost when fire strikes in those tall buildings. Firefighters are confronted with extreme circumstances on every high rise fire. Indianapolis has a separate operations guide specific to high rise firefighting. It is our HOG (high rise operations guide) manual issued to each firefighter, one of the recommendations of the panel. We pray that another tragedy like that won't happen to us again.

The aftermath of that fire continues to haunt our department. Gary, my classmate who helped save the downed firefighter's life, later retired due to a medical disability. Recently and tragically he took his own life. The suicide and alcoholism rate among firefighters is higher than the general population. We are tortured souls.

Chapter 26

The good old chief's aide retired. His wife threw a big retirement party for him. I was sad to see him go. He often discussed with me how much he hated that my classmate badmouthed me at other companies in the battalion. He said he went behind him and cleared up the stories. One thing he questioned, would I, or any woman, be able to do the job when we got into our 50s? If he could only see me now, he would certainly agree that I am still quite capable of firefighting, although, like my old captain, everything is stiff when I get up from a chair.

Our new backstepper and the aide's replacement arrived. We had ping pong tournaments again. The lieutenant and I kicked butt as usual, but these guys, good sports, took it in stride. No more paddles were thrown after a loss.

The three shifts decided to make some major renovations to the firehouse in addition to our annual spring cleaning. We wanted to add some rooms to the unfinished basement area, mainly a workout space and a bathroom. For months, every duty day, we came in and did construction for eight hours or more, in between runs and cooking. It interfered with my workouts. I had to wait until evenings to run and lift weights, which sometimes, too tired, I skipped. I was investing in real estate on my days off, old HUD repossessed houses, which needed renovating. I got tired of doing construction all the time. I wanted a break when I came to the station. I was extremely happy when we finally finished our construction project.

The lieutenant was a great guy, but he could be terribly moody. He put everyone on edge when he came to work in one of his black moods. We nicknamed him "Anus", because he could be such an asshole on those days. I had my bad days too. The guys called me Sybil because they said I had many different personalities, and they never knew who was going to show up for work. My other nickname was Leona, because I was collecting rental houses. The lieutenant meant it in a nice way, but I was pretty sure the backstepper called me Leona because he thought I could be a bitch like the hotel heiress.

We had this whole secret code going that we would use on the street. Number 1, “Have a nice day”, translated was “Fuck you”. Number 2, “Good morning”, meant “Kiss my ass”. We made a list with the translations and assigned them numbers. In the mornings we would tell the lieutenant number one, number five or number twenty, but we meant it with love.

Late one night, while the lieutenant was on vacation, the three of us, the two backsteppers and I, were dispatched to investigate a report of smoke in the area. Most of the time, these calls were bogus and a waste of time. We got to the area where the smoke was reported and slowed down. We could see a slight haze and smelled something burning. A garage fire was reported a couple of streets over. Some other companies were dispatched there. Not seeing anything significant, we slowly headed in the direction of the reported garage fire. Down a side street, I thought I saw a glow as we drove by. I backed up; then we all saw it. Two adjacent houses were on fire. The one house was rolling, fully involved and caught the one next to it on fire. We turned down the street, stopping to catch a plug. The backstepper, in the lieutenant’s seat, was yelling into the radio for additional companies. He yelled at me to stop in front of the fire.

I said we should pull up, but he said, “Stop right now!”

He jumped out, grabbed a line and started putting water on the fire. I dropped the tank for him (Dropping the tank is using water from the engine which carries 500 gallons. This gives me, the engineer, about five minutes to get the line hooked up from the hydrant before running out of water.) I got the hose clamp on the water supply line (normally the lieutenant’s job) and dragged the line to get it hooked to the engine. The new guy went to help the backstepper. When I climbed up to the pump panel, I saw that I could barely keep one line supplied from the hydrant. The water pressure was so low that I would burn up the pumps if I pushed any more water. I knew we were in trouble. The heat from the fire was cooking my engine. It would only be a few more minutes before the paint started to blister. The yellow, plastic "Keep Back" cushion over the end of the ladder was melting. I grabbed my trusty, red line and sprayed the engine to cool it.

You are probably wondering why I just didn't move the engine. Once pumping operations are started, the engine can't be moved because it is out of motion gear and in pumping gear. If the engine has to be moved, the water must be shut down first, the gear changed, the engine moved, and then the water can be pumped again. It is important to properly position the engine in the beginning because moving it later is almost impossible.

Other companies arrived but I couldn't allow anyone to hook up to me since I would run out of water and burn up the pump. The chief sent some to look for more hydrants since I couldn't support any more lines. They had the same problem, proving it wasn't incompetence on my part, which I was sure some people were thinking. Finally, a call was made to the water company who upped the pressure in the water lines. That fixed the problem but didn't help much because the one house only had a foundation left and the other one was a total loss even though it was still standing.

My engine had to have the "Keep Back" cushion (a bright, yellow, padded, cushion with the "Keep Back" warning written in large letters on it which serves the purpose of warning drivers to stay back from the engine but also protects our heads when we run into the feet of the ladder which hang beyond the back of the engine) replaced for which I took a ribbing. To no avail, I tried to tell everyone my backstepper made me stop. The chief, always on my side, was the only one who believed me. We discovered the reason the water pressure was so low in the area was because kids kept turning on hydrants due to the heat. They'd play and cool off in the water until they got chased off. To combat the loss of water, the water company lowered the area pressure. It worked fine until the fire. Most likely, we would have saved all but the siding on the house next door and some of the first house, if we had just had regular hydrant pressure.

A 911 call came early one morning. A neighbor woke up to find a kid, probably 18 or 19, hanging by a rope from a tree. We woke his household to tell them there was nothing we could do for him. He was very cold, his face purple and his body ashen gray. He had been hanging for awhile before we got there. The police arrived and wanted to make sure murder wasn't involved. The mother, of course, was screaming and crying. From the neighbors, we discovered that he had attempted suicide twice before. The father was distraught but able to talk. The police took pictures even though they believed it was a suicide. There was no box or chair by him. It was deduced that he jumped out of the tree, making it a sure bet his neck would snap. He was serious about making this attempt successful. In the meantime, we had to listen to the mother begging for him to be cut down. We were released from the scene by the police and the ambulance crew. I was glad to get out of there because I sympathized deeply with that poor mother's crying. My nephew by marriage hung himself. This tragedy happened a few doors down from where our backstepper lived. He decided to move.

Some people in this area didn't take baths or showers, except maybe on Saturday nights. If you met them on a Friday, they were very odiferous. We were dispatched on a late, night run for an unconscious person. We got there and were escorted to a bedroom where an elderly woman was not breathing and pulseless, naked on her bed. The room was crammed with boxes and papers leaving only a small path open to the bed. Since she was still warm, we needed to do CPR on her, but it can't be done on a bed. A solid base for compressions on the chest is needed or there won't be good blood circulation. The two of us each took an arm and the lieutenant took her legs around the knees. We carried her to the living room, struggling because she was a big lady and because we had to hold her up higher than the bundles we were squeezing through. It seems to be a rule: the bigger the person, the further they need to be carried. The lieutenant made awful faces, and his eyes were watering. We laid her on the living room floor. I moved to her chest to start compressions. I realized why the lieutenant was close to gagging. The vaginal and anal smell coming off this woman was putrid. Her armpits weren't nice either, but the other end was practically unbearable.

The lieutenant said he would be right back and ran off. I wondered if he couldn't handle the smell. He came back quickly with a towel and stuffed it between the lady's legs. The ambulance crew arrived and started an IV, giving her meds to jumpstart her heart. They put a breathing tube down her throat. We loaded her onto the cot and took her to the hospital with the towel in place. The ambulance crew thanked us for the help and the towel. Unfortunately, the lady couldn't be revived and was declared dead at the hospital. We laughed at our own sick jokes which were about her being buried with that towel between her legs and her inability to throw anything away. We needled the lieutenant because he could not handle the smell, asking him if he needed roses to be able to do his job.

The lieutenant and I argued over many things. He was my boss, but we had to live together, so frequently I voiced my opinion. Being a cold natured female, we battled over the temperature in the firehouse, especially in the summer when the thermostat was set at freezing. In the winter, on the way to runs, the lieutenant rolled down his window because I turned up the heat in the engine. The control was on my side, the vent was on his side. He threw his coat over the vent to stop the heat from coming out. I turned it up full blast negating the blocking effect of his coat. He rolled down his window.

We fought over the radio. The fire department had installed a new radio system, and we could shut the radio off during the day. The alert would make the radio come on automatically whenever we were being dispatched. He liked it on all the time so he could stay aware of what was going on in the city. I constantly shut it off for peace and quiet. I reasoned if a second or third alarm (great big fires) was going on somewhere, control would make their special alert for the whole fire department, therefore, I did not consider it necessary to ruin my TV show or phone conversations with a constant blasting of a radio in the background. Ninety percent of the fire department feels the same way I do, but my crew teased that all I cared about was a paycheck. I countered that they were eaten up and didn't have a life beside this job.

Sometimes the lieutenant and I got into fierce, all-out combat, but since I couldn't get too mouthy, I found sneaky ways to get him back. He knew I checked the engine and washed it first thing every morning. If something was wrong, I reported it to him. If nothing was wrong, I went to do my workout in our newly, renovated basement without making a report.

Usually he was fine with this, but on one of his cranky mornings, he hit me on the intercom asking for a report. I was 15 minutes into my exercise-bicycle ride, so I ignored him, hoping I could get the last 15 minutes in before he noticed I hadn't answered. Five minutes went by and he called for me again.

The chief's aide, not intentionally being a busy body but trying to help out, stuck his head in the door and said, "Your lieutenant is calling for you."

I couldn't pretend any more that I didn't hear the announcement. I had to run up to the watch room where he was sitting at the desk doing paperwork. He chewed me out for not giving him an engine report and delaying coming. I declared I hadn't heard him so he told me to keep the stereo turned off while I worked out so I would not miss a run. I said I wouldn't miss a run because lights come on with the radio alert. He threatened to write me up for insubordination. He was in a bad enough mood to do it, too.

I shut my mouth and said, "Yes, sir."

It was my day to cook. Nobody cooked with pepper because the lieutenant had a hiatal hernia and was prone to indigestion. I was making chicken and noodles with mashed potatoes and green beans. I loaded everything with pepper. The lieutenant loved my chicken and noodles, and ate a bunch of food, then suffered the whole night. I caught him unaware. Subsequently he tested his food anytime we had a fight. The new guy and the backstepper said we fought like we were married. I swore many a time I was going to divorce him.

The chief and I exercised together, speed walking around the block. We talked about everything under the sun. One day, we were engrossed in conversation and neglected to listen to our radios, both of us carrying one as we strutted. As we rounded the corner, half a block from the firehouse, the engine, being driven by my backstepper, pulled up with lights rotating and sirens blaring. Shouting out the window for me to get in, I asked what was going on. The lieutenant said we had a run. Not sure if it was more shenanigans, I looked at my radio and the chief. He was checking his radio. The lieutenant yelled to get in because we needed to go. My ass would have been chewed for an hour for missing the dispatch over the radio but since the chief had been with me and hadn't heard the run either and the lieutenant couldn't chew his ass, I never heard a word.

On a run, the lieutenant smarted off to me in front of some civilians. I was embarrassed by his overbearing meddling and got mad. I purposely put the patient information sheet on his side of the dash. It was normally kept on my side of the vehicle. On the way back, I got rolling fast, then asked him to hand me the paper. He had to lean way up to reach for it. Just as he got his hand on it, I slammed on the brakes and he smashed his nose on the windshield. I told him I had to brake for a squirrel. He told everyone I was trying to kill him. That wasn't true. I just wanted to hurt him a little.

He swore he would get even. He brought in a plastic, baseball bat and whenever I had to create G forces due to a wrong turn, he beat me with the bat. The backsteppers loved our wars. They were a constant source of entertainment for them.

Sometimes the fun got out of hand, and I was usually the victim since it was three against one. A sub came to our station and burped in my ear after a good meal. It was a loud, obnoxious belch and I let him know I didn't appreciate it. It then became common practice for my crew to burp in my ears since I didn't like it. I couldn't rip burps like them. Remembering how they ran from the front porch spitters, I warned them I would spit on them if they kept burping on me. After one too many burps, I spit at one of the guys who jumped back missing the wet fusillade. Finally, they took me seriously.

Nobody burped in my ear until a week later, when the lieutenant, missing the fun of my disgruntled reactions, fired one off. I warned him but he let an even louder one off, and I spit, hitting him directly in the face. Infuriated, he tried to grab me but I leaped from my chair and ran off, heading for the refuge of my locker room with him hot on my tail. I slipped and fell going around one of the corners after pushing through the first swinging door of the hallway to the locker room. The lieutenant pounced on top of me, pinning me down. He started spitting in my face. I spit back, both of us landing some good shots. Since he is germaphobic, he couldn't handle much spit getting on him. He gave up quickly, got up and ran off to wash his face.

I went to my locker room, washed my face, fixed my ponytail and went back to my chair in the watch room. The lieutenant, still seething, enlisted help to pay me back. Gladly, enthusiastically, his two back steppers volunteered. They snuck up on me, armed with first aid tape and Kerlix, a gauze we use to wrap wounds. They held me down, taping my arms together, my legs together, and securing me to the chair. Then they left me, stuck in the chair. Their laughter turned to silence as they scurried off. I was asking them, begging them, shouting at them to please let me up. I sat quietly for a while. I wasn't too uncomfortable since it was a nice rocker recliner chair. Pretty soon, the tape started itching and I needed to pee. I yelled for help. The chief came back from his travels in the district and heard my cries. He hurried in and attempted to unravel me. Being too difficult to loosen the tape and gauze by hand, he got some scissors out of his office and cut me free.

When I was a sub, and jumped on the beds of the regular guys who were trying to sleep, I was threatened with being taped to the hydrant out front. As they were dragging me away, they let me go because of my pitiful pleading. No amount of begging, crying or pleading stopped my heartless crew from taping me up. The chief was my knight in shining armor, protecting me from my own crew. They said I was just a kiss ass. I said they were misogynists picking on the lone female. I peppered the food heavily that night when I cooked. The lieutenant went hungry.

We were dispatched on a fire just a block from the engine house. It was a big church and all of the firefighters needed multiple tanks of air to get it under control. I was pumping a lot of water to multiple lines, coming off both sides of my engine. We were using all fog nozzles (a nozzle that has a wide pattern of spray rather than a straight stream) so I didn't have to adjust the pumping pressures for each hand line. I was completely set up before anyone came out for the first exchange of tanks. My crew came out. The lieutenant came to me bending over so I could put on his new bottle of air. Somebody handed me a tank, and I asked if it was full. Yes was the answer. The lieutenant told me to hurry so I didn't double check to see if the tank was indeed full. I screwed on the new bottle, and he ran back into the fire.

A few minutes later he was back, yelling at me that I could have really killed him by giving him an empty air bottle like that. He said he was deep into the building when he took his last breath. He said I tricked him good because the tank was so low on air, it was beyond activating the low-on-air-indicator which is a vibration of the facepiece when the air is turned on. I swore to him I would never mess with him like that, and it was purely accidental. I denied wanting him dead. I liked playing the game for fun, but not for real. To make up to him, I cooked his favorite dinner without pepper. When we reminisce, he swears I almost killed him that day.

Big night fires are murder on kidneys. If I wake up in the night at the firehouse, which I usually do two to three times a night, I get up and go to the bathroom. At times, when I have been asleep for three to four hours and awakened by the alarm, there is no time to pee before jumping on the engine.

At 0200, a fire came in. The first arriving company didn't lay a line, common in those days because many fires could be put out with the 500 gallons in the engine tank. They reported heavy fire in one house with two exposures. Extra companies were called. We came in on the original dispatch and saw a plug as we approached from the opposite direction. My lieutenant ordered me to hook up the big five inch hose directly to the hydrant; then my crew dragged two, 3 inch lines to the engine in front of the fire, hooking my engine to that engine.

Our department had just moved from 2½ inch hose to 3 inch lines. The couplings were still 2½ inch negating any changes to the hydrants, and we still would fit other local department connections. The increase in water flow and decrease in friction loss gave us the capability of pumping a lot more water out of our engines which, in a situation like this, gave us a great advantage. I pumped from my engine through those three inch lines giving the engine at the front of the fire a ton of water. The middle house had fire coming out of every opening. The windows were blasted out with the flames shooting across, catching the houses on both sides on fire.

My crew ran up with their line and hooked up to the engine into which I was pumping and disappeared into the house. The standard procedure for a back up crew is to assist the first in crew. The crew in the house asked my crew to get them something and they refused. Instead, they went around them to grab the glory and the fire. Other arriving companies were sent to the exposures to cool them down with a protecting stream of water, preventing further damage to them from the house that was still rolling with fire despite the internal opposing attack.

I was left by myself, half a block away from the fire, to tend to my engine while it pumped. After a few hours, my eyeballs were floating. The fire was out, but there is always hours of work afterward. Holding my pee was becoming painful. I saw a lot of the guys disappear behind buildings. I figured it was time for me to disappear somewhere too. Spotting two houses close together, I walked between them and stood for a while making sure the coast was clear. Pulling down my pants, I squatted to pee. It felt heavenly to let that vice grip loose and free the dammed waters. Just then, a big dog came running around the back of the house, barking, snarling, jumping on the fence making the chain links rattle. I fell over from fright, and peed all over my pants. I got up and finished peeing, keeping a close eye on the dog as it barked its head off. I was glad the owner of the house didn't come out to see what all the commotion was and catch me with my pants down. I pulled up my pants, buckling them as I ran back to the engine. With my heart still thumping, I checked all my gauges to make sure everything was okay. The pee spot on my pants got cold and itchy. We were there another two hours. Daylight was peeping through by the time we got back, and I headed straight for the locker room to pee again and change pants. The hot shower felt good, relieving the itch.

It was my lieutenant's turn to get an ass whipping from the chief. Since the chief and I had become great friends, I hung out in his office with him and his aide in the evenings. We talked about our families, vacations, told jokes and watched TV together. We didn't hash over fire department events preferring to talk about other things.

The day after the triple house fire, my lieutenant was sitting in his office. I came in to give him my engine report. He asked me what I thought about the situation with the first company complaining about them on the fire. I said I didn't know what he was talking about.

He said, "You mean the chief didn't tell you about it."

I said, "No, we don't talk about the department."

The lieutenant, not really believing me, explained anyway. The company they went around on the last fire, called the chief and complained. The officer said it was a constant problem with the guys at my station. They were always trying to hog the fire, no matter who was first in. Instead of assisting the first in company, they freelanced and did whatever they wanted, never what they were asked. The chief gave my lieutenant a lecture about being a team player and not a Prima Donna. The lieutenant was despondent that other firefighters would feel this way about him. He took the constructive criticism aiming to be better, which I think he accomplished.

Chapter 27

Rumor reached us that headquarters was planning for a second hazardous materials team to be placed somewhere in the city. My lieutenant reasoned that if trained people existed at our station, the chief of the department would take a closer look at us for the team. The lieutenant asked each one of us if we would volunteer for hazmat training. I was all for it and so were the backsteppers. I definitely believe if you can get paid to get an education, do it! I applied for quite a few schools offered by the department in the few years that I had been on but was never picked. I was hoping, with the lieutenant's backing, I would be accepted. A few workdays later, we got the news. We were all to be trained as hazmat technicians, the highest level, and it would be on duty.

I can't remember how long the training lasted, but it was typical, fire department, company training. We did some learning, some joking, and some cat napping. Class was eight hours each duty day. We had long, lunch breaks and usually let out early. The material was technical in nature and hard to stay awake for, especially after eating. The blasting air conditioning froze me. The shivering kept me from dozing off most of the time. I brought a sweatshirt every day, but my fingers still felt like ice.

The class clown (there are usually several in every fire department class) brought in a watch that had VCR controls on it. The instructor couldn't figure out why the tape kept pausing. He gave us a break calling for a technician to look at it. She said it was working fine. When he resumed class and it acted up, he got suspicious. Everyone denied messing with him. Defeated, the instructor gave up on the VCR and had slideshows instead.

Test question answers became obvious. The instructor would say, "Pay close attention to this slide. You might see this on a test."

With hints like that, it's hard not to pass.

After the written test, we moved on to the practical portion, which was staged in a huge warehouse at the shops. We learned to plug leaks with: putty, wooden plugs, oversized containers, and so on. We got into suits that made us look like we were ready for a trip into outer space. (These were the same suits that we wore when we picked up the dime in the academy.) For fun, we pushed each other down, which wasn't difficult because the suits are so oversized, they made us klutzy. When someone suited up goes down, it's like a turtle on its back unable to get up without help. The new guy and I spent more time knocking each other over, than anything else. For training, it was pretty fun. The main drawback to the practical training was the heat. It was summer and the warehouse was hot as hell. Putting on the suits added to the heat so we drank a lot of sports drinks. Since we were a young crew, all of us under 40 (some of us just barely), we could handle the heat. But I was always glad to get back to the air conditioned firehouse where we could battle over the thermostat setting.

After class, not far from the firehouse, we got a run to a residence for an accidental poisoning. Enroute, we were advised the child was actively seizing. Around here, it might mean something terrible or it could be a faker trying desperately to get some attention. We arrived to hear a lot of shouting and panicked bustling. A mom struggled to put a spoon in her daughter's clenched mouth. The girl was a cutie with red, curly hair and a few freckles on her cheeks. She was in serious trouble. Her head was in a rhythmic back and forth shaking pattern. The girl's eyes were shaking too. Mom stated she is normally a perfectly, healthy child, no history of seizures. Her brother was the one who had seizures. His seizure medicine had been in the bathroom cabinet. Mom came back from shopping to find her daughter like this with the empty bottle of seizure medicine on the floor.

The babysitter, a 14-year-old neighbor, had no idea how the girl got the medicine or how long ago she had swallowed it. Mom said she called 911 right away and while she waited, the girl's eyes rolled back in her head and she swallowed her tongue. She heard somewhere a spoon should be put in the mouth to hold the tongue down. I gave her a lecture, never put hard objects in the mouth because of the damage that could be done to the teeth, airway, soft palate, and tongue. She said she was scared because she didn't want her daughter to choke or stop breathing and she was trying to help. I told her she could make things worse. I was trying to be instructional, not scolding.

When the medic arrived, she grabbed the girl and hurried with the mom to the ambulance. They sped off with lights flashing and sirens blasting. Ambulances seldom go to the hospital in emergency mode because 80 percent of the time, it isn't really an emergency. We knew the situation was serious by the actions of the ambulance crew.

When I saw the medic on a run the next duty day, I asked how the little, carrot-topped, girl was doing. She died. I thought of how I treated that poor mom, telling her she made the situation worse by shoving the spoon in her daughter's mouth. I probably added tenfold to her grief. She was untrained and scared to death for her daughter but she took action and I lectured her. I imagined her sitting at the funeral home, crying her eyes out because she did the wrong thing. I felt terrible because I made the situation worse. I never saw that mom again, but if I did, I would beg for her forgiveness.

I totally forgot the feelings of the grieving, horrified, family members another time. It was twilight and we were sent to a street directly behind the firehouse. The neighbors heard some gunfire and called 911. We arrived to find a 20-year-old male lying in a puddle of blood. His young wife, with their baby in her arms, was screaming and crying, leaning over her dead husband, rocking back and forth on squatted legs.

She was sobbing, "He went to get milk for the baby. We needed milk for our baby."

He was shot when he rounded the corner. The milk broke open when he dropped it. He walked part of the way, holding his hand to his throat, dripping blood as he crawled the rest of the way making it up the steps to the storm door where his bloody handprint smeared down the glass as he collapsed on the porch.

We got out the BVM (bag valve mask) to pump air into his lungs since he wasn't breathing. We beseeched his wife to move away so we could work on him. She moved but was still in the way. We asked again, but she didn't move. I yelled at her while glaring until she finally backed away. My lieutenant pulled on my pant leg to refocus my attention on my job which was starting CPR. I dropped down to my knees and squeezed the BVM to give him his first breath. The air came bubbling back out the bloody hole in his throat, rattling the tissues and exposing the internal flesh. A white piece of something was stuck in the wound. The backstepper and I fingered it in order to discern if it was the windpipe cartilage. The backstepper finally recognized that it was wadding from a shotgun shell. He said the gun had to have been held right at his neck when he was blasted because that's the only way he could have gotten wadding embedded in his wound. There must have been several people holding him while the gun was rammed in his neck and fired. This guy had some serious enemies.

He was a nice looking, young man with blond hair and blue eyes. The little baby, who wouldn't get to know his dad, was as cute as his father. Most likely it was a drug deal gone horribly bad.

When the ambulance crew got there, they attempted to get a breathing tube down his throat, but couldn't get past all the wadding and pellets. We picked him up, threw him on the cot and raced off to the hospital, doing CPR as the medic loaded him up with heart medicines. He was pronounced dead at the hospital. We could hear his wife's wails as she was informed that nothing further could be done. My actions haunt me. That poor girl was in shock at the sight of her dead husband sprawled on the front porch. She was frozen, looking at him, unable to take her teary eyes off him, while I was shouting at her to move. I should have gently moved her. She left the hospital, only to arrive home and see the bloody porch. She had to wash off the clotted, congealed, blood of her husband's spilled guts, the remaining stain hounding her.

As a crew, sometimes we needed a break from the gore of the streets. The new guy's family had a house on a lake. He wanted to start having a party every year where we could swim, fish, boat and ski together with our families. Our schedules were hectic, but we finally settled on a day where everybody thought they could make it. We played cards with his mom and dad, drank beer, and went out on the boat. He pulled my kids around the lake in a tube hooked to the back of the boat. They had a blast. It was a small, shallow lake. Hundreds of fish jumped out of the muddy water from under the turning propeller. We laughed as we watched the shiny, silver backed, white bellied, little fish sparkle in the sun. We stayed all day swimming and boating, leaving with our stomachs full of hot dogs and hamburgers.

The department mandates EMS training frequently in order to meet state standards. At these training sessions, we practice all of our skills, especially the ones we don't do very often on the street. We were scheduled with another company, but they got a fire before marking out of service. That left just my crew and me. The instructor had us practice all kinds of splints on each other. We used a long traction splint specifically designed for upper leg fractures of the femur. It is important to reduce or pull the overlapping bones apart to keep the muscles from cramping in the thigh. It lessens the pain, bleeding, and likelihood of nerve damage. It's a complex process, so we practiced several times. It is amazing when it is put on victims, usually from auto accidents, who scream while it is being done, but breathe a sigh of relief after it is set.

Since we seldom backboard victims who are standing, we practiced that too. The instructor wanted one last volunteer to be tied to the backboard before we went to lunch. Everybody pointed at me. I told him I didn't trust them after the chair debacle but they sweet talked me and promised they wouldn't abandon me. Being the sucker that I am, I let them secure me to the backboard. The instructor had them turn me over to prove a patient could be secured to a backboard in any position. Then they asked the instructor about this position and that position, flipping me all around, up and over, head over heels until I begged to be let down. At the instructor's behest, they gently laid me down, still tied to the board and left for lunch. After a few minutes, the instructor came back and untied me saying the joke had gone far enough. Letting bygones be bygones, we all went out to lunch together at a restaurant of my choosing.

One of the guys who teased me a lot, for example, he would say I was only on the job for a paycheck, called in because he was going to be late to work. It was 0750, and he had just awakened. We were supposed to be at work at 0755. Nobody said much because he is a big baby if anyone teases him. He was the one who threw the ping-pong paddle after losing. When he made a second big mistake in a row, everybody laid it on thick.

We got one of our requisite night runs. Since this was a busy engine house with three or four runs every night, everybody was used to getting up. Sometimes, I was somewhat foggy and wouldn't be sure of the address. On this run, I got in to drive and asked the lieutenant for the address but he wasn't sure either. I got out, ran into the watch room, tore off the printer sheet with the address, and got back into the cab. We hauled ass out of there.

Normally, I check the back to make sure everybody is seated and ready but I forgot since my routine was changed. The lieutenant didn't look back either. We arrived at the address. As we approached, the ambulance crew waved us off before we got out of the cab. We were overjoyed and thanked them for not needing us. We drove back to the firehouse and saw someone sitting on our bench out front. Who would sit there at 0300? Sometimes people walk up to the firehouse for help. As I backed in, he approached the apparatus. He was not a sick person. He was our backstepper! In disbelief, the lieutenant and I double checked behind me. Nope, he wasn't there. How did he get out there? As soon as I parked and climbed out, he yelled at me for leaving without him. I went back to bed. He stomped around, throwing a tantrum, complaining to the lieutenant as I shut the bedroom door. I heard the lieutenant tell him to get his ass out of bed next time because he had plenty of time to catch the engine.

The next morning, at the kitchen table, the chief said, "I was awakened by a bunch of yelling last night and came out of my quarters to investigate. Your backstepper was in his underwear running out the overhead door, chasing after the engine as it pulled away."

We routinely shut the door with a remote unit as we leave, securing the firehouse. If we forget or for some reason the door opener/closer doesn't work, we call dispatch and ask the police to come by and close the door for us. The backstepper forgot all about the door coming down. After he realized he couldn't catch the engine, he stopped in the street, then heard the door closing. He turned and ran to beat it before it shut. The door won. Locked outside in his underwear he pounded on the door to get back in.

The chief let him suffer for a few minutes, teasing him, "Who is that? What is wrong? What are you doing out there?" then opened the door for him.

He put on his nightpants and waited on the bench for our return, pouting and cursing.

After that, when he got phone calls, he was paged by every body as part-time firefighter or volunteer firefighter who comes in when he wants and goes on runs when he wants. His face was bright red for weeks. I'm sure his blood pressure skyrocketed every time he was paged. The new guy told him he should lay off the caffeine slurries before he had a heart attack. His slurry was a cup of coffee with two scoops of hot chocolate mix, caffeine loaded. We told him to patent his recipe.

Somewhere around this time, a female captain applied for and was put in charge of the training academy. It was not a position any other chiefs or captains wanted. She challenged the haircut policy for women by telling the new female recruit not to cut her hair. Hearing about her bold initiative, I quit cutting my hair. It grew fast and was past my shoulders in no time.

At my station we had a run early in the day while I was in the shower. Jumping out, dripping all the way to the engine, I caught the run with wet, stringy hair. The incident was a house on fire, and several chiefs were paged to the scene. One of them saw my stringy hair which was almost dry by now. He chastised me for not having my hair within regulations. When my chief heard him chewing my ass, he interrupted the complaining chief and took him off to the side. Hearing snippets of the conversation, my chief basically told him that he knew nothing about the situation and he would handle the matter. He never bothered me about it but I am careful to always wear my hair up on runs.

Chapter 28

On Sundays, I loved to get the real estate section of the newspaper and peruse the listings, searching for bargain houses in my area of town. I discussed, at the firehouse, how I found several steals using that method. Sundays are nice because most firehouses consider it a day of rest. There is no extra housework and everybody sits around chatting longer than usual. Suddenly, I could not find the real estate section anymore. Mysteriously, no one knew anything about its disappearance. Even the chief was clueless. He or his aide could usually point me in the direction of the culprit. For years, there was no real estate section.

After a few Sundays of searching, I solved the problem by bringing in the real estate section from my home and looking it over in the women's locker room. I didn't tell anyone because I figured it would disappear from there too. I had my suspicions. I thought it might be the new guy because he was a trouble maker or in fire department lingo, a shit-stirrer.

Just in case it was him, after the backstepper woke us every morning by shoving his bedclothes into a crinkly, plastic, garbage bag, I got up and turned the new guy's mattress upside down with him in it. Then I would run into the refuge of my locker room.

The next duty day, first thing in the morning, he would usually push me around as a payback. Most times I just ignored it unless he was particularly obnoxious, then I pushed back and it would turn into an all out wrestling match. It got to be an every day occurrence. We were fairly well matched, about the same size and weight. I wrestled with my older brother growing up. I knew a few moves. I would get him in a scissors lock with my legs and squeeze him until he would beg for air with his last breath. He got me in a headlock and pushed me to the ground.

If my mother had been there, she would've said, "Stop it before someone gets hurt. When you horse-ass around like that, somebody will get hurt and I don't want you to come crying to me when it happens."

I rose up with my back against the weight of his body. A loud crunching sound came from my back and an electrical current shot through me. He sprang off me when he heard the crunch, asking if I was okay. I told him I thought something was seriously wrong because it hurt like hell. He ran to get the lieutenant. The chief was gone in the district.

My three crew members came, hovering over me, asking me questions. The lieutenant wanted to know what happened. I told him I was running through the bay and slipped in the water, fell on my butt and heard a loud pop in my back. He said he fell on his butt off a ladder one time while painting a house. It stunned him and he couldn't move. He thought if I sat there for a while, maybe I would feel better. The backstepper got a chair and the three of them helped me up so I wouldn't be on the cold, cement floor. After 10 minutes, with no improvement, the lieutenant called for an ambulance.

They walked me to the cot and laid me down. I was in a great deal of pain, shivering all over. The EMT packed towels and blankets under my back to support it which alleviated the shooting pain, leaving me with a sharp pain that magnified with every heart beat. Taking off for the hospital, we hit bumps. With every bump, pain shot through my body. Ambulance rides are very bumpy.

Arriving at the hospital, I was wheeled here and there, ending up outside the x-ray room. I was lying quietly, and finally, with four blankets piled on top of me, stopped shivering. Some doctor came and by himself, rolled me over to feel my back. Generally, it takes three people to move a back/neck injury patient. I cried out in pain. He dropped me quickly and ran off. I started shivering again, as the pain shot up and down my body. He must have said something because immediately after that I was taken into x-ray. Gently, they slid films under my back leaving me there while they were developed. I heard a rush of people come in and lots of whispering. A doctor informed me I had fractured a vertebra, T-12.

Understanding what he said, I squalled, "Oh God, I broke my back!"

The doctor, trying to console me said, "It's in the flexible part of your back so it won't affect your range of motion too much. You are lucky, because it doesn't appear to have damaged your spinal cord."

I didn't feel lucky at the moment.

As usual, I had to pee. Some things never change. The nurses were ordered to keep me still and wouldn't roll me or lift me to pee. They finally, with permission from a doctor, inserted a catheter into my bladder to drain the fluid which filled almost half a bag.

The nurse said, "Wow, you really did have to pee!"

The doctor poked me all over to make sure I hadn't lost any neurological function. To the right of the broken vertebra, it was tingly and weird feeling as he poked there. Even now that area is hypersensitive to touch.

I called my older sister, who is a nurse, wanting her to gently break the news to my parents. Next, I called the firehouse to tell them. The chief, whose brother broke T-12 after a fall off a ladder and was wheelchair-bound, was shaken by the news. They all came to the hospital to see me, bringing lots of candy, as soon as they got off duty in the morning. I was heavily sedated and strapped to the bed so I wouldn't move and make the injury worse. A man came to measure me for a body cast which he said would take a few days to make. My lieutenant apologized. He said he should have called for an advanced life support, paramedic ambulance. I told him not to worry because they wouldn't have transported me any differently.

I nibbled on the candy but was feeling too sick to eat. This really alarmed my crew. Katie (another nickname) won't eat candy! She must be near death! The doctor said it is normal to feel nauseated with head and back injuries. It is the body's way of telling a person to be still, lest they complicate the problem. Laying absolutely still caused gas to collect in my intestines blowing them up like those long balloons clowns twist into animal shapes. The pain and cramps from the bloating became as severe as the back pain. The nurse gave me stool softeners and medicine to break up the gas which helped.

Firefighters from the union came to visit, along with my personal friends and family. My lieutenant came every day. I knew he had better things to do. I told him he didn't have to come all the time, but he did anyway.

On the sixth day, my custom cast arrived. It spanned my entire back, from my armpits to my butt. It was carefully placed on me after I was given an extra injection of painkiller. Now, I wanted to go home. Before the hospital released me, I had to prove I could pee, poop, walk, and switch to oral pain medicine. I am great at peeing, even after being catheterized for a week. I was told that some people lose function in their bladder after catheterization. Not me. I hadn't pooped all week, but I also had not eaten. I wasn't sure I could muster a bowel movement. I quit complaining about pain.

First things first, I had to sit up, which proved to be the most difficult part. The nurse raised the head of my bed to get me used to sitting up. Every hour they raised it a little more. My chief came by and sat for hours with me, wiping the sweat off my forehead as waves of nausea raced through me. Finally, I raised my head off the pillows and let my legs dangle as I rested my head on the chief's shoulder. I wondered if the chief was gagging over my body odor. I hadn't showered all week, only sponge bathed. My hair was greasy and matted to my head. All the sweating, with no deodorant, was formidable for my armpits which were getting pretty hairy. Now that I had quit taking shots of pain killer and was receiving only oral pain pills, I began to think and care about my hygiene. I didn't want a towel stuck between my legs. The chief, being the kind man that he is, suppressed any revulsion he might have felt and encouraged me until I was finally able to plant my feet on the floor and stand up. My legs were rubbery, but they held.

A physical therapist showed me how to use a walker. I caught on fast, even the tricky part of going up and down stairs. I walked very slowly, but had no problems with balance on my shaky legs and arms. I lost some weight but still had no appetite. I forced myself to eat shepherd pie and with the exercise finally worked up the release of a small turd at which the nurse clapped. I begged to be released. They advised me to stay a few more days. I said I wanted to go home ASAP. I explained that my sister is a nurse and she could look after me at home. They reluctantly agreed.

On the seventh day, somewhat biblically, I was sent home to rest and recuperate. Getting out of my sister's Cadillac, my transportation home, the kids jumped up and down with glee at my arrival. I felt very sick from the ride and took medicine to calm the waves of nausea. Heading straight to bed, I fell asleep from exhaustion. The kids woke me with boos and scary noises, ready to go trick or treating, dressed in their Halloween costumes. With hurried kisses, they ran off to fill their bags with a booty of candy.

The next day, my sister helped me get up to pee. I had a portable potty by my bed so I only needed to take a few steps. My stomach did flip-flops, but I traversed the short distance. After I lay down, I felt a few pangs of hunger. I ate a grilled cheese sandwich, then delved into all of the candy my friends brought me and snuck some of the kid's Halloween candy. Thereafter, I was in a feeding frenzy. For months, I laid in bed, eating out of boredom, packing on the pounds.

Finally, the doctor said I could walk in the neighborhood on my walker. I started with slow, short walks, but by the third month, I was race walking on my walker for miles. One of my friends, an old college classmate, lived in the neighborhood and I walked to her house frequently to say hello. The first time she saw me, I thought she was going to cry. She got used to it and walked with me a few times. After three months, the cast came off and I got to take a shower. That was heaven.

The calls and visits from work stopped. It was as if they forgot me. My niece came to live with me, so my sister could have a break. My niece liked living with my family and me, and stayed long after I was capable of taking care of myself. I felt reasonably well until one day I sneezed and almost passed out from the pain.

The doctor said it was time to start physical therapy. Therapists are trained in torture and killing. They don't care if you're in pain and struggling. They push you until you reach their goals. My doctor was the chief of neurosurgery. The fire department was going all out, getting top doctors and great therapists. On a visit to the doctor's office, I asked how long he thought it would be before I could return to duty. He gave me a stern lecture about banishing that thought from my head. He said I would never return because the next time I had an accident like this I would probably end up in a wheelchair. I was told my bone mass was good and it was just a freak accident. Why couldn't I return? How likely was it to happen again? I told him firefighting was in my blood, and I couldn't possibly do anything else. He was adamant about protecting me like my lieutenant was adamant about protecting those kids. I started crying.

Leaving his office, I went to the fire department doctor whose office was on a different floor. She saw me right away and calmed me down. I begged her to give me a chance to get back to firefighting. As a woman doctor in a field dominated by men, she understood my conviction and vowed to help by speaking with the neurosurgeon.

The next month, when I saw the chief of neurosurgery, he said I could try to make it back but he doubted it would happen. He swore it would be like training for the Olympics and winning gold. I told him I was up for the challenge. I rode the elevator downstairs and thanked my fire department doctor for her help.

I intensified my physical therapy, feeling my blood pressure shoot up and getting headaches while riding the stationary bicycle. The worst pain happened while doing twisting motion exercises. I fought through the excruciating pain but it was exhausting. I went home and rested for hours. I stopped gaining weight but didn't lose any either. I weighed about 170 pounds. Prior to the accident, I was a comfortable 140 pounds.

Every year, the Firefighters Credit Union has a dinner and dance to honor us, their heroes of firefighting. Our station decided to attend as a unit. I had been to a few of the dances and was looking forward to it. The first time I went was right after I was divorced. As I deposited money in the credit union, the president of the bank asked me if I was attending. I told her I couldn't afford it. Knowing my banking information, she knew I was telling the truth. She escorted me back to her office. I was shocked when she tried to give me a ticket for free. They were $30. I refused, but she insisted. I had a lot of fun and went frequently after my finances were in order.

For this dance, I wore a black, sequined dress with a stretchy waistband. It was the fifth month after my accident. I stayed for a couple of hours, danced a few slow dances, shook some hands, and got very tired. I chatted with one of the guys involved in a fire where a wall collapsed, falling on top of him, breaking his back. He was trapped in the debris with his arms and legs pinned, running out of air, thinking he would die because he couldn't breathe. He said he moved his head back and forth pushing the mask off enough to get some of the smoky air in his lungs. The smoke made him cough, but at least he was able to breathe until they dug him out. He was carried out on a backboard, unable to walk out on his own. Due to his lower back fracture, he could walk only with leg braces and had to retire. He works in communications as a civilian. If I didn't get busy, I was going to end up a civilian too.

Monday, after the credit union dance, my phone rang early. It was a chief at headquarters telling me I had to report to the fire department doctor for an evaluation to come back to light-duty. I was not ready to sit in a chair for eight hours a day. The doctor agreed with me and asked why I was stupid enough to go dancing in front of the chiefs. I said it was only slow dancing, and I was exhausted the next day. She bought me another month of at-home recuperation, but then I had to report to the shops for desk work.

I felt fairly sick the first week, having to take frequent breaks to rest my back. But what doesn't kill you makes you stronger. I was too tired the first week to go to therapy. I struggled to make it to therapy the second week. After a month, I was back to pushing hard at therapy. The guys from my station came to the shops for repairs to the engine. I hadn't seen them since my release from the hospital. I was stuffed into my pants barely able to buckle my belt. The wretch who broke my back could not help but notice my increased girth and rode my ass.

He said, "Don't get too close to her because that button on her pants is ready to fly. With the force behind it, when it blows, it could put out an eye."

So much for any sympathy for my struggles!

All this work was becoming a pain in the back, literally. I went to the doctor complaining about increased pain. She sent me for an MRI or magnetic resonance image, which clearly shows soft tissue injury. Two discs in my back were bulging, causing the pain. I was sent home for three weeks and cried about the setback. I saw the images on the screen of the computer as they were printed for the doctor. My back looked like a train wreck. I wondered if I really would make it back to regular duty. I stopped by the firehouse, confiding in Ruth about my doubts. She said she was betting on me and cheered me up a little. She told me not to get impatient or lose heart.

I returned to light duty and physical therapy, invigorated after my rest. I graduated from general therapy to specific therapy for my occupation. I stopped by the firehouse to see my lieutenant and he asked if I was ever going to make it back. I told him I was working on it. I asked him why he never called after I left the hospital. He said he didn't want to pester me or disturb me at home. I told him it made me feel unwanted. He apologized.

At about that time, the secretary to the chief of the department, whom everybody loved, died suddenly at an early age from a brain aneurysm. She was married to a retired firefighter. We all grieved for the loss of her and for her poor, brokenhearted husband. It was the second marriage for both of them, but a match made in heaven. The chief needed a person to fill in for her until he could hire a permanent replacement. His aide called me at the shops, saying they needed a hard-working, dependable person for the chief. Would I be willing to come downtown to the City-County building and work at the chief's side?

I was at the shops doing all the crap work nobody else wanted to do. I caught up files that were months behind. I worked hard for the secretary, thereby making her job easier. She did me favors in return. During spring break, when the kids were out of school, she let me bring mine to work nullifying the need to pay a babysitter. There was a conference room with a TV and a big table where the kids could play. We ate TV dinner lunches together. I stayed in the conference room with my kids, answering the phones, which didn't ring much since most of the office was also gone on vacation.

How could I leave someone like that for a chief who kept me from promotion for years now? Maybe he wasn't the one who "lost" my grievances, cheating me out of promotion points, but I was sure he was involved in some way. For the past three years, since he came into office, he could have promoted me at any time. If he and his aide thought I was a hard worker and dependable, why wasn't I already promoted. Based on this, I declined, telling him I wouldn't be a good candidate because I would be gone for therapy too much. Perhaps that was a mistake. If I had gone there, the chief would have gotten to know the sweet, lovable person that I am and maybe he would have promoted me, then again, maybe not. I would have brown-nosed for nothing. We will never know.

After 14 months, the final, two months filled with grinding, physical therapy, I was sent back to the doctor for a definitive evaluation. I was now under the care of the fire department doctor, the nice woman who fought for me. She gave the okay to the physical therapist to put me through a timed agility test. She strapped a leather brace around my back, similar to what weightlifters wear. I loaded four, not the normal three, hose lines over my shoulder by lifting them off the ground and flinging them up on my shoulder. They made me carry extra hose because I wasn't wearing my gear; coat, boots, and night pants. I had on an SCBA or I probably would have had to carry five hose lines. I felt it was overkill, but I guess they wanted to be sure I could still do the job. I went up and down four flights of stairs, dropped the hose pack, and simulated: opening a hydrant, pulling down a ceiling, raising a ladder, swinging an axe, and lifting a cot. My therapist said I lifted the cot with more force than most of the men who came through. I passed. I won the gold.

The therapist sent me to my doctor with the results. The doctor, pleased with the report, congratulated me and said I was a true, warrior-woman, which the cook, who had depicted me as one in his acid etching, knew all along. She released me for duty. I drove directly to headquarters reporting to the chief of personnel, who released me to the companies without performing the fire department physical agility test. He said the hospital's test was more difficult than ours, so there was no need to delay my return. I was elated.

Heading back to the shops, I said my good-byes which weren't tearful because I would see everybody whenever the engine needed repairs. Reporting to the firehouse the next day, I proudly handed my lieutenant a back to work slip from the doctor. I couldn't have been happier. The guys said it was good I returned. They didn't have anybody to pick on while I was away, and had to turn on each other. Now they could focus on one target-me-which they did with a vengeance.

The backstepper who was supposed to stay away from caffeine was back to drinking slurries and grouchier than hell. The new guy, not so new anymore, let his "little man syndrome" increase to giant proportions. He teased me endlessly about being fat. The lieutenant, as usual, stood by and let us establish the pecking order. All of this in-house bickering didn't bother me until they started insulting me in front of civilians. I got mad and asked for it to stop. It only increased. These guys, if I let them know something bothered me, it seemed to excite them and add fuel to the fire, if you will forgive that expression.

The problems were compounded when I brought in some laundry from home. I decided to donate some of my old, too-small, clothes to Goodwill. I wanted to wash them before I gave them away. Some faded Polaroid pictures that must have been hidden away for years, dropped out of a pocket onto the floor as I was loading the clothes into the washer. I didn't notice them, but the next person using the washer, the grumpy backstepper, found them. They were clear enough to see my naked body in sexy poses. These were some pictures my ex-husband had taken of me, many years before. The backstepper showed them to me at arm's length, and then ran off with them, hiding them. I snuck into the locker room that night to retrieve them. Since I had been gone, the station was robbed two more times, so all of the lockers were locked.

Now he started pestering me for sex. When I did not oblige, he got belligerent and nasty. I treated him with a cold shoulder, which pissed him off even more. The lieutenant asked me why the backstepper and I were horn-locked, but I couldn't divulge the problem. Instead, I told him I was tired of all the bullshit from every body. I wanted it to stop, especially on the street, in front of civilians and ambulance crews.

Soon after my return, we got a nasty fire where a mom was badly burned trying to rescue her three kids. She was battling for life in the hospital. My crew did an excellent job of putting out the fire but was too late to rescue the unfortunate little kids who burned to death in the inferno.

They hauled the children out back, so the media and public couldn't see the little bodies. My crew tricked me into going in the backyard, where I saw their dreadfully shaped, charred bodies. Their legs were drawn tightly against their butts. Their bony arms were outstretched as if waiting for someone to come pick them up. Their faces were stretched into grimaces, exposing all of their teeth. Their eyes bulged forward with the lids burned off. The skin on their bodies was pulled taut, making them look like skeletons with a leather layer, more grotesquely shaped than mummies.

They were lined up on their backs, in the order of their size, a two year old, four year old, and a six year old. It turned my stomach, but I was hardened enough that the smell and sight didn't make me puke or pass out. I think the guys were disappointed. They were hoping for a big reaction.

Like a little boy holding up a dead mouse to scare his sister, the guys often tried to get a rise out of me. These are dark games we play. Our work often requires us to rub elbows with destruction and death, warping our sense of humor. Although this behavior may seem strange to people outside the realm of firefighting, this twisted, outlandish humor is an excellent coping mechanism. When discussing our sick humor with retired friends, they reassure me that I will revert to normal after retirement. That's a surprise since I wasn't normal to begin with.

Part Five

Out of the Pan and Into the Fire

Chapter 29

My young, neighborhood friends who had crushes on me returned for a visit, dropped off by their parents. They had grown considerably in the time I hadn't been able to see them. They welcomed me back with a card and candy. The lieutenant said after a couple of visits, they could not come back anymore, even though their parents were bringing them. I protested. The lieutenant said the grumpy backstepper complained about them playing with his gear and was afraid they might give him lice. This set a bad precedent. Nobody else was limited by who could visit. The new guy's kids came frequently and did homework at the office desk.

I asked for help from the chief, but he said he couldn't interfere with the lieutenant's decision since he isn't much more than a guest in the house. This was more than I could take. I requested a move. The chief begged me not to go but I was ready. I wanted to go back to a truck, where I would go in to fight fires again instead of staying behind to pump the water for everybody else. I missed the excitement and challenge of, as one chief aptly described it, "slaying the dragon".

I wasn't the only unhappy person. The grumpy backstepper also put in for a move. If I had known he was leaving, I might have stayed. After five years of being there, having arrived on the same day, we both left on the same day. I told the lieutenant that our divorce was granted and I would see him on the flip side. I was moving to a west side truck. I was apprehensive because I didn't know the west side worth a diddly. I would have to learn the district all over again. It was sad because I had finally stopped mixing up East and State Streets.

I bought a boat, and since it was summer, I received permission from the parents of one of the boys who had a crush on me to take him to the lake with my family for a day. He was a good kid, on the honor roll in school, and I wanted to do something special for him. He had never been on a boat and was timid. By the end of the day, he was splashing around in the water like a fish, just like my kids, but buoyed by a life jacket.

I arrived at my new firehouse to find it was in an uproar. There was only one backstepper on the truck and he fought constantly with the captain. The captain didn't know the district very well so he couldn't help me with directions. The engine crew couldn't stand the truck crew and vice versa. The other shifts were sometimes drawn into the controversies. There were some bright spots. We had a permanent cook, the truck backstepper. The house was recently remodeled so I wouldn't be facing any big construction projects at work and I knew the house captain. She was trying to make chief and was the victor in the long-hair-for-women campaign.

The fire department was co-staffing ambulances. I bid for a position and got a downtown ambulance stationed at a firehouse. I would be assigned there once a month for a 12 hour shift, reaping a bonus of $50 for riding out. The guys were a bunch of partiers at that station. I had a lot of fun with them. We played cards, Parcheesi, basketball, and generally goofed around. At night, they snuck women in. They made regular visits to a newsstand where they purchased fine cigars and girlie magazines which they took back to their locker room. Police officers were always dropping in to join the activities. A bakery was across the street which delivered boxes of day-old doughnuts. Everyone was very nice to me. It was a relaxing change to come here from my station where everybody butted heads. Plus, I kept up on my EMS skills.

The chief of our district, who was stationed at another firehouse, was nice but didn't like our captain. My captain deserved to be yelled at. He would go into fires, swinging his arms. He didn't bring any tools like the pike pole or axe. He often didn't get on his gear, standing outside, directing people. That was the chief's job. The captain was supposed to throw ladders, chop holes, pull ceilings, and knock open doors for engine crews so they could get water on the fire.

Trucks get more fire than engines because they serve a bigger district. Eighty percent of our runs are EMS and engines are sent out first for EMS. Trucks serve as back-up EMS. Since trucks don't have many runs, I got to sleep through the night frequently. We did the truck roll when the engine went on its nightly EMS runs. After getting up several times every night for five years, it was a refreshing break to not be exhausted after most duty days.

One of the first fires at my new house was in an apartment complex. Our engine was on a first aid run. My truck responded and the first engine company was coming from a good distance. The captain was on vacation. We had two subs. We pulled up with heavy fire showing from the second floor. The two subs raced downstairs and searched for victims. The acting officer and I hoisted a ladder and went up on the roof with a chainsaw. An engine and another truck arrived as we were cutting a large vent hole in the roof. Two from the other truck company joined us on the roof. We cut across a 2 x 4 truss, so the square we cut didn't drop. I stomped on it with my foot and down it went. As I wobbled, my partner grabbed me, so I didn't go down with it. He said I gave him a scare and I should be more careful because he might not be able to catch me the next time. The board crashed through to the living room because the fire had burned through the ceiling. The smoke billowed out and I could see the engine crew making their way in.

We climbed down the ladder and joined the engine crew inside. We had successfully vented the place which would greatly help the engine crew do their job since they could see much better. They were making excellent progress putting out the fire. Our quick, fire stop on a potentially volatile apartment fire made the captain look bad because we did a better job without him. Our engine missed the whole fire because of the medical run, which made them extremely upset. They cursed EMS.

We had a dryer fire at a laundromat in a strip mall that smoked up the place and the businesses around it. As usual, my captain came in, swinging his arms. I brought in a pike pole and an axe. Both were needed. The chief noticed I brought them and knew how to use them. He pulled me aside after the fire was out and the smoke was cleared from the buildings. He told me I didn't have much leadership, and liked the way I took initiative anyway. He said the captain was being counseled, and I should keep up the good work, because there was a possibility my captain would be gone soon. He wanted me to continue my hard work because if I took the captain's lead I would have to be disciplined too.

We woke up one night to the report of a warehouse fire. We zoomed out in our usual order of engine then truck. We heard enroute that it was a fully involved mechanical warehouse. We were ordered to set up our aerial on the west side of the building. The captain told the backstepper to get on the aerial ladder and ride it up as I extended it to save time. I told the captain it was extremely unsafe to operate the aerial that way. The captain took over the controls and rode him up. When I went to engineer school, we were sternly lectured about the safe operation of an aerial. Ten years before, a firefighter had his feet smashed in the aerial rungs as he rode the ladder while it was being extended. He had to retire. Several years after that, another firefighter had a foot smashed in the rungs. It was a big no-no to do that.

I reported the captain's actions to the engine captain, who was also the house captain, on the other shift. She wrote an entire report with a list of complaints from everyone. Our cook had put in a move because he and the captain had an argument over the cook's motorcycle. The cook loved his motorbike and babied that thing, shining it, cleaning it and storing it inside. When he rode it to work he parked it behind our truck where the captain liked to sit watching TV and chain-smoking. The captain, in an effort to quit, wore a nicotine patch, but still smoked. He claimed it helped him cut back. The cook, polishing his bike, whistling and moving around behind the captain, irritated him because he didn't like the commotion around him. He ordered the cook to take the bike outside and leave it there. The cook refused. The captain threatened to write him up for insubordination. The cook told him to go ahead because he didn't care. The captain got him laid off without pay. The cook, an old-timer, put in his move, got it and left. Everybody was furious with the captain because no one should mess with the house cook.

With everybody ganging up on him, he was moved out of the companies to be evaluated for fitness for duty. He put in a move to a north side company and got it. The office, always trying to support its officers, let him transfer for a fresh start in another company. Perhaps he would learn from his mistakes. With him gone, all of our vacancies filled fast. We got a nice captain who came from the office, freshly promoted. The truck acquired a young, hulking backstepper and a skinny backstepper. Two new subs, one of whom was female, went regular and chose the back step of our engine.

One of the engine crew members roofs houses on his days off. I had a couple of roofs on my rentals that needed replaced. I dreaded the big bucks I would have to spend to hire a roofing company. Mentioning the projects to him, he said he would help but with certain stipulations. He hates tearing off old roofs and carrying shingles up the ladder. If I did those two things myself, he would give me a great deal on laying the shingles. I paid my brother to help with the tear off. Considering we didn't know what we were doing, we did a passing job. It is hard work, but simplistic. The most important thing is having a good, roofing shovel to peel back the old, crumbling shingles and roll them down the roof.

Since I climb on roofs regularly for my job as a firefighter, I was comfortable spending all day perched up there working. The engine guy and I have done a lot of roofs together now. He brags, to my embarrassment, that I, his little lady, can keep up with and outwork most men.

My secret is consistency. I pace myself and can generally plod along all day. Another big secret, (Don't tell my roofer. I'd hate to spoil my reputation with him.) I have a crane load the shingles on the roof. It costs a little extra, but it's worth it. Who wants to carry all those heavy, floppy bundles up a ladder? If my roofer doesn't, I sure as hell don't want to either.

We frequently ran with a township fire department whenever there was a run dispatched at the fringe of our district. If they needed help, they would call for us. (IFD runs in with all of the township departments surrounding the old city and they run in with us. It's called mutual aid.) We had a fire with them in some nice apartments, a high value district. When we arrived, they had most of the fire out. Preparing to make a secondary search for victims and check for hot spots, as soon as we got close enough to enter the place, before putting on our masks, we could smell the gasoline. Some arsonists are stupid. Gas has a distinct odor, making it easy to tell when it has been used. The arsonist obviously didn't care because he was trying to cover up the murder he committed. In the middle of the front room was a partially burned body. When we rolled the body over, the stab wounds were apparent. We left the body in place. Arson was already on the way. Now homicide was alerted.

Most of the residents, awakened by the sirens and flashing lights, were crowding around to get a glimpse of the body. The press arrived with cameras rolling and pushed their way through the shocked residents. Since they monitor our radio chatter and know all of our codes, it is hard to keep any secrets from the news stations. The press was out in force, like ants at a picnic, getting in places they are not wanted. With just a few police there, and the rest on the way, it became a challenge to finish putting out the fire. It seems that in nice areas of town, murders are much more sensationalized because of their infrequency. Unless somebody squawks in the poor areas of the city, they are not much more than a footnote.

When I was at this station for about a year, the administration announced a new promotion process. The list, with my name still at the top, was being thrown out. I called and asked if there was any way they would consider promoting me before abandoning the list. The answer was an adamant no. At this point, even though it wasn't a surprise, I mourned the loss of my deserved promotion. For six years I sat at the top of that list. Contacting a lawyer again with my sad tale, I was given the same answer, without proof, I didn't have a case. To the disappointment of the chief at my old house, who called to console me, I decided not to enter the next process. He said every process has a next-on-the-list who didn't make it. It was not soothing rhetoric. I declined, deciding to put my energy where I couldn't be thwarted by cheating despoilers. I refinanced my houses and with the extra money, bought a few more rentals.

The fire department paid us overtime since the inception of the federal Fair Labor Standards Act. We got a big paycheck every third week because we worked 72 hours one week a month. Firefighters had to be compensated with overtime pay any time we worked more than 48 hours in a week. The city didn't like paying us overtime. The union, with the backing of the chief, negotiated a Kelly Day for us in lieu of pay. Every seventh shift day, which is every 21 calendar days, we are given a day off. If you look at it another way, approximately once a month we get five calendar days off in a row. It is very sweet and I don't know how we lived without Kelly days. There were a few who griped because they wanted the overtime checks. About 16 or 17 years before I came on, firefighters worked 24 hours on and 24 hours off. They were given an extra day off once a month, also called the Kelly Day, which gave them three days off in a row. Those old guys worked a bunch and dirt cheap. Today, we are better compensated for the work we do.

With the new Kelly Day, we were assigned workgroups of nine or ten people. I was fortunate to be at a station where I was fourth in seniority with only 12 years under my belt. I would get a Saturday or Sunday Kelly Day and some nice summer vacation picks. I was pleased to get, in my opinion, the sweetest Kelly Day, Saturday. I would not work a Saturday for a whole year.

The new system meant that the least senior person would fill in for Kelly Day people or if no one was on Kelly Day or sick or on vacation at the station, he would rove (move to another station in the district). The least senior guy at single engine companies might have to rove frequently if all vacancies are filled. Firefighters could have 20 plus years on but still have to rove if they are at a house filled with very senior people.

Another change occurred; we used to take vacations based on department wide seniority. Now, we picked in house first and department wide second. Since Fridays, Saturdays, and Sundays were already taken up by Kelly Days, we could only get vacation days Monday through Thursday on in house picks. I took all the summer weekdays that I could get. I wanted off while the kids were out of school. I had seven vacation days to spend which is equal to about three weeks of vacation. Some of the less senior guys got mad because the captain and I took big chunks of the summer vacation leaving little for the ones behind us. We senior people told them, "Get some time on the job!" (Any gripe based on seniority was usually answered that way.)

Even though our two engine backsteppers were from the same class, she had better scores in school and graduated higher in class. Lower in class ranking, he had to rove whenever we had two full crews, which wasn't often. At some stations, backsteppers took turns roving but she refused. She felt she earned her position and was not giving it up. She, rightfully so, told him to get some time on the job.

With our fresh crew came a new aerial. It was a beauty. She was prepiped. We could skip the time consuming process of running the big 5 inch hose up the ladder. In comparison to the old aerial, it was quick and easy to set up. On an experimental basis, it was equipped with air conditioning. The whole department was jealous; at least, I thought they should be. At this time, trucks were outfitted with big cutting tools and spreaders used for motor vehicle accidents. After receiving our training on the special equipment, we took them out and practiced frequently. When we needed to utilize them to free trapped vehicle accident victims, we looked like the experts we were paid to be.

We had a run one night where a trapped victim was fading fast. The medic got in the vehicle to start an IV while we worked to free the patient. By the time the paramedic got the IV going, we had the driver's door cut off and the roof peeled back allowing total access to the patient and the ability to get a backboard cleanly in the vehicle thereby protecting the victim's spinal cord. The medic was impressed.

The truck rarely had EMS runs, but I admired the knowledge of medics and wanted to become one. I was riding out on the ambulance once a month and was told I should become a paramedic since I incessantly asked questions. A program was announced, and I applied.

We were given study material to prepare for an entrance exam. They assigned each applicant a day to test basic medical skills. Even though I shouldn't have been, I was nervous. I performed well until I came to the CPR station where I forgot to give extra breaths, failing to hyperventilate the patient. (We no longer do that. It has been found harmful.) It was called a critical skill and automatically eliminated me from the process. I was sorely disappointed and was told it was a shame, because everybody thought I would make a good medic. The rules were clear, though. I was encouraged to practice and apply the next time a class was announced, which would not be for another year or two. Since the criteria changed each time, just like the promotion process, it was difficult to know what to study.

We wanted some new equipment for our weight room. Pestering the house captain to buy us some, she said there wasn't enough money to purchase good equipment.

Being a good house captain, she devised a plan, seeking donations from the community. Each shift went to designated businesses in the neighborhood, asking them to pledge money. We hit the jackpot when we approached a franchised gym which had just purchased some new equipment. As the new equipment arrived, they donated the old stuff to us. Headquarters okayed the donation. After a few months, we had a great workout room.

I believe there were eight machines for working different muscle groups and a treadmill with incline functions. I wore myself out on all those machines, loving every minute of it. A couple of us worked out together. It was embarrassing to lift weights after our hulking backstepper. The rest of us had to remove a bunch of weights from the barbell in order to make it human lifting capacity. He had his weakness though, multiple surgeries on his knees throughout his years of playing football. He said his knees were bothering him again and he thought he might need surgery soon.

I beat him to knee surgery. It was my turn to be on the ambulance. After getting multiple runs without making it back to the firehouse, the ambulance was finally pulling into the station. Because I had to pee and was about to bust my bladder, I opened the side door and jumped out, intending to run into the bathroom before we got another run. The ambulance made a wide arc before backing in. The force of the turn must have transitioned into my knee because it twisted as I hit the ground, buckling under me. I heard a loud pop that reminded me of the sound my other knee made years earlier during a basketball team practice when I completely tore the anterior cruciate ligament (ACL) in my left knee. (That injury almost kept me off the fire department. I had to go through some extra steps in the physical to prove I would make a good firefighter in spite of a knee prone to hyperextension.) The pain was excruciating, but I knew I had to crawl away from the backing ambulance or get run over. The ambulance crew saw me inching along on the ground in their mirrors and stopped. They helped me hobble into the firehouse on one leg.

The acting lieutenant called the safety officer who came in his van to take me to the emergency room. When the safety officer arrived, the acting lieutenant wouldn't let me hobble to the vehicle. He picked me up, carrying me in his arms. I am no lightweight, weighing about 160 pounds. I was chauffeured down the street a few blocks to a clinic where my knee was x-rayed. No broken bones were found but I was not satisfied. I've had plenty of strains and sprains in my life, and I knew this was more than that. Even though I pleaded, the doctors said I needed time and wouldn't consent to an expensive MRI. Given crutches and a soft cast that spanned the length of my leg, from my ankle to my hip, the doctors sent me home for two weeks rest.

My knee ballooned to twice its size, even though I stayed off it, keeping it iced and elevated. I was sent back to the shops after another two weeks when my knee was more normal looking and I could drive a car again. My knee swelled any time I was on it. The doctors decided I might need an MRI due to possible soft tissue injury damage. DUHHH!!! The MRI showed a shredded anterior cruciate ligament, a common but devastating injury in women athletes. The specialist said I should retire. I did not like this negative quack.

Calling other firefighters who had knee surgery, I asked for recommendations. It appeared the most experienced doctor around was the team physician for our professional football team. He was just who I needed, a doctor used to dealing with sport injuries. Luckily, he was in the insurance network our city used. If I had the best surgeon operate on my knee, I hoped I would be able to stay on the department.

Three months into my injury, my knee stopped swelling, if I was careful. My new doctor had no question that I could go back to the fire department, if I underwent surgery. Encouraged by his positive attitude, I asked if he could build a new ACL in my basketball injured leg. He said it was a possibility and that he had performed double ACL replacement surgeries for several other patients.

After he x-rayed it, he said he wouldn't touch it, "It looks like a war zone in there. If I was to tighten it up, it could be permanently painful because you have no cartilage left. It is bone on bone. Since you got along with it for 20 years, let's leave it that way."

I was disappointed because my knee is not stable when I jump. I can't compete for rebounds when playing basketball. I was hoping for a miracle cure. So much for getting two cures with one surgery.

My surgery was scheduled early in the morning-the doctor's first one of the day-catching him when he is fresh and eager. I was nervous before going under the knife because I would be knocked out. If something goes terribly wrong, it is usually from the anesthesia. Anesthesiologists are the most sued doctors. As I was waiting for the anesthesiologist to give me his knock-out drugs, I stifled a panic attack. I told myself, you can do this; just stay calm; it will all be over soon. I wanted to jump up and run out of there. If I did, I would never be able to return to firefighting so I gutted it out. At last, the anesthesiologist had my IV started and wanted me to count backward from 100. I think I made it to 98.

When I awoke, freezing cold, my knee was killing me. The nurse gave me more blankets and administered pain medicine saying she thought my doctor never used enough. I was glad she was sympathetic because the shot put me right back to sleep, glorious sleep, where one feels no pain. I was dimly aware of being moved to a hospital bed. Insurance wouldn't cover a 24 hour stay at the hospital for the surgery. I stayed overnight, 23 hours, leaving early in the morning.

The doctor ordered me to walk on my leg from the hospital bed to the wheelchair without any crutches. He said my knee was repaired and it positively would hold. I had to trust him and my knee. Except for the gut-rending pain, when I placed my weight on it, everything was just fine. They encouraged me to walk as normally as possible to the wheelchair. But when you feel like you are walking on hot coals, it is tough to be casual. I smiled at the doctors observing me as I sat down. They nodded their approval as I was wheeled to my car. A special machine was loaded into the car. As soon as I got home, I popped some pain pills and hooked up my machine. Strapping my leg into the bars of this torture device, I perched on the couch and hit the on button. It kept my leg in constant motion, bending it at the knee, back and forth. This, I was told, would keep scar tissue to a minimum.

I spent two weeks on the couch, literally tied to that machine. The kids got themselves off to school with bowls of cereal for breakfast. The whirring of the machine was irritating. To combat the noise, I put on headsets and listened to the radio. A special ice pack on my knee kept it cool, minimizing the swelling and subduing the pain. I imagined I was on a well deserved vacation with time to laze around and read books or watch TV. Since this was a line of duty injury (no matter how stupid), I was being paid to lie around and recover. Of course, I wished none of this had happened, but being an optimist, I try to look on the bright side of things.

After four weeks of recovery at home, it was time to go back to light duty and time to meet the torturers who call themselves physical therapists. Having gone through physical therapy for my back, I was not looking forward to it. To return to the firehouse though, I had no choice. It would have to be done. I was allowed to go back to my familiar stomping ground, the shop.

The chief initially assigned me to work the front desk at headquarters. This is crappy duty. Nobody wants to work the front desk, being tied to it, bored and answering phones all day, not allowed to leave without relief, even to pee. At the shops, you're free to pee whenever. I called the secretary to see if she would make a request of the chief to assign me to work with her. I only had to beg a teensy-weensy bit. She knew I would work hard for her.

For the next three months I was at the shops and in therapy. Some new vehicles were ordered and arrived for the chiefs. I drove them to different businesses to have custom features installed: sirens and lights, brackets for SCBA's, emergency radios, etc.

A retired captain oversaw the work the city garage performs on the fire apparatuses. Firefighters constantly gripe about the work done at the shops. The mechanics complain about the firefighters disregard for daily maintenance. The retired captain mediated.

Sometimes the captain took the secretary and me out to lunch for a treat and to thank us for our good work. Since we were watching our weight, most of the time we ate TV dinners together in the conference room. Because weight jumps on me when I am not exercising at full capacity, I couldn't afford very many restaurant visits where I tend to over eat.

For Christmas I was given big cans of popcorn as presents from everyone. Since I had cut out candy, I popped microwave popcorn for snacks, and it didn't go unnoticed. The shop treats their disabled firefighters well.

Working five days a week 8:30 a.m. to 4:30 p.m. is taxing if you are used to working 24 hours then having 48 hours off. It's somewhat like having a weekend off every other day. I had to farm out some repairs on my rentals because I didn't have my normal days off to make repairs. I am very handy, fixing most of the problems at my rentals myself. On the other hand, it can be lonely being away from your family on work nights since their schedule is five, eight hour, work days or school days, and weekends off. Then again, sometimes it's nice to get away from the whining and griping that happens every night in families with kids. Being gone on the weekends is the worst when you have to leave everybody at home still sleeping in their beds. At least I got weekends off working this schedule.

A lot of firefighters call in sick on Saturdays and Sundays. Recently, I saw a vacation schedule in a chief's office. Fifteen people are allowed to take vacation days through the week days. Only five people are allowed vacation days on Saturdays and Sundays because the office knows that many extra people will call in sick. Slackers have all the fun.

The city required the department to inventory items at the firehouses. Being available, I was sent with the list to check off items accounted for and to add anything not counted. I went into every room at every firehouse, escorted into some of the men's locker rooms. At some houses, I was directed to knock on the men's locker room doors and then let myself in. The difference between the men's furnished locker rooms and the women's bare locker rooms was glaring and downright pitiful. For the most part, the women's quarters had no benches or chairs to sit on, no telephones, no clocks, no radios, no weight scales, no nothing. I brought this to the attention of several people in the office, who pooh-poohed it. I made sure the chief got the news.

When he wouldn't respond, I called my old buddies at the local chapter of the National Organization for Women (NOW). They felt this was a matter for the state president. She arranged a meeting with the chief which Ruth and I attended. Ruth was elected to a national committee looking over minority rights at fire departments across the country. We lambasted the chief with the differences. The NOW president pointed out the men had a place to sit down to change their pants. Did he expect the women to hop on one leg while changing?

Ruth brought up the security of the locker rooms, especially when no women were regular or permanently stationed there. She told the chief she roved through a station where feces were smeared on the walls of the women's locker room. The chief, demanding to know what station, excused himself and went into the next room. He called the captain at that house. We could hear him yelling as we waited. Returning red faced, he closed the meeting promising he would remedy the situation.

Even though he was the chief who wouldn't promote me, he kept his word on this. New locks were installed on all of the women's quarters. Matching keys were allocated to all of the women and house captains. The rooms were furnished with chairs and benches. Phones were installed. All of the officers were warned to keep them locked, and the men were to be kept out, or there would be hell to pay. The men grumbled that their extra shower was being taken away but that was quickly squelched.

Chapter 30

Released by my doctor, after brutal physical therapy, I returned to work. Again, I did not have to take the physical agility test to prove I was fit for duty. All smiles, I was happy to go back. Conversing with Mary, she was miffed that I was never required to take any agility tests when returning to work after injuries. She had to take an agility test after each of her pregnancies and wondered what was up with that. I sympathized with her.

Several hours after getting my aerial washed and shiny, we were dispatched to the report of a fire. We saw the smoke in the sky as we rounded the corner. Pulling directly in front of the house, we laddered it as the engine crew went inside to put out the fire. Due to the steep roof, we carried an extra ladder that has hooks on the end which can be unfolded. Bringing it up the ladder we climbed, we laid the hooks over the ridge of the roof. Scrambling up the hook ladder, hanging off it with our tiptoes in the rungs, we chopped holes in the roof to release the smoke and gases. It took a lot of strength to get through the shingles and the old 1 x 6's with axes. I was exhausted after what seemed to be an eternity of hanging off the ladder at a precarious angle. We got the job done. Hot black smoke came billowing past our faces causing tears to well up in our eyes and snot to run from our noses.

Barely able to come down, the realization that I came back too soon after my injury stunned me. I promised myself I would spend more time on our fancy, donated equipment to rebuild the strength in my legs and arms. Shimmying down the ladder, I neared the last rung on my wobbly legs with shaking arms. The axe slipped out of my hand and clattered off the roof to the ground below. The chief's aide was standing close by and picked it up for me.

He handed it back, giving me the evil eye, as I finished my descent off the ladder. We were both grateful that no one got clunked on the head by my falling axe. Several times I have seen axes fly out of people's hands. I saw a captain get knocked in the helmet with one. One time, I was torpedoed by a falling axe.

My aerial was dispatched to a fire on some new construction. We put out the fire quickly, since there was no furniture, clothes or other obstructions in our way. It burned the roof badly, and the boys from another company were up there chopping away to make sure the burnt areas were cut out to prevent a rekindle. My captain and I were looking out a hole where a window was going to be placed, inspecting the damage and determining if we needed to take further action. We heard some yelling above, then THUNK, an axe hit my extended arm! It hurt like hell, giving me a muscle cramp that wouldn't quit. The firefighter who let it slip was lying on the roof looking through the hole saying sorry and asking if I was okay. The axe hit me with the back of the blade, so I did not have a chop out of my arm, just a charlie horse that was assuaged with a couple of ice bags from our medical supply kit.

The captain, just like in the academy, sat me down for the rest of the cleanup. He wanted to send me home, but I refused to go declaring I just needed a little time until it stopped cramping. I worked out the pain by flexing it and moving it around. The medic massaged it and gave me aspirin for the pain. We carry aspirin for people with chest pain in case they are having a heart attack. Chewing six of them helped me too.

By the next morning, my entire forearm was black and blue. It could've been worse. The axe could have hit me in the face. One of the women caught a falling ladder in the face. It knocked out some of her teeth, besides giving her one hell of a concussion. She retired on disability.

Speaking of concussions, I have received two. We were sent to help a diabetic. He was very combative, and we needed help holding him down. We asked for police assistance. That gave us five people to secure him and one person to start the IV and give him some glucose. Where diabetics get the energy to wrestle and squirm away from us, since they are low on sugar, is perplexing. Our patient was middle aged, not burly, but strong as hell and crazy wild. He, of course, was not aware of what he was doing and had no idea why all these people were surrounding him. We finally got all of his limbs and head locked against the floor.

Popping out below the tourniquet, the medic saw a promising vein on which we were concentrating. Our man pulled an arm free, swinging a fist that landed squarely on the side of my head, knocking loose my grip from his other arm and messing up the IV. We had to wrestle him back down as we listened to the medic spit out a stream of curses. The man's wife was in the background crying, asking us not to hurt him. Never mind that I was the one with the black eye developing. As we wrestled him down the final time with secure holds, the sugar was put directly into his bloodstream.

He immediately quieted down. After a minute, he was back to his senses. It is miraculous how fast the glucose works. Someone told him to look at my eye where he had punched me. He cried and apologized, swearing he would rather cut off his arm than do that. He begged for forgiveness. My pain and his sorrow made a tear roll down my cheek. Accepting his apology, I left to go sit in the ambulance with an ice pack over my eye while the medic got the man to sign a paper which stated he didn't want to go to the hospital and releases us from liability if he should die because we didn't transport him.

Most diabetics don't go in an ambulance to the hospital because they have already done that many times and know how expensive it is. Generally, they need a good meal and a visit to their doctor. We reminded him to eat and waved good-bye after he signed. Some diabetics would have to go to the hospital with us every other week if they couldn't sign an SOR or statement of release. I went to the hospital for my throbbing head later.

The other time I got a concussion was a similar situation. We were wrestling a big guy who drank a fifth of whiskey while downing some pills. It started when his girlfriend couldn't rouse him. She got scared and called 911. The drugs were illegal and she delayed calling until hysteria took over.

When we got there she was frantically crying over and over, "Please don't let him die! Oh God, don't let him die!"

Never mind that he shouldn't have taken illegal drugs and she should have called earlier. He downed a narcotic, which slowed his respiratory system. The alcohol he drank is also a sedative which equates to a double dose of downers making him a very sleepy man, too sleepy to even breathe.

The medic got a line started to give him a drug that would reverse the effects of the narcotics and allow him to breathe on his own. We breathed for him until the IV medicine took effect. One important thing the field textbook highlights is to give just enough medicine to get the patient breathing at a healthy rate. Do not give too much or the patient will wake. Our medic gave a pinch too much and woke him. He was scared and pissed, probably delusional. Let the games begin.

We got him by all fours, securing him. The police had arrived long before due to the overdose dispatch and they helped hold him. We were tying him to a backboard when one of his legs got free. He kicked me in the head, knocking me backward over the couch.

Now the girlfriend was screaming, "Don't kill him! Don't kill him!"

The cops screamed at the guy after he knocked me over, cuffing him to the cot. As we wheeled him out, the police asked if I would like to press assault and battery charges. I declined. I was sure the guy would be sorry that he kicked me. He wasn't himself and would most likely be ashamed when he came to his senses.

I arrived at the firehouse with my head reeling. After an hour or two, I went to the bathroom because I thought I was going to puke. I lay down on the cool floor resting my throbbing head against the tile. A run came in, but I couldn't get up. The medic burst into the locker room to get me for the run. Finding me on the floor, she yelled for help. They contacted dispatch to start another ambulance replacing them on their run, in order to load me into the back of theirs. They drove across the street to the hospital. I got a CAT scan which confirmed a concussion, was given some medicine to squelch the nausea, and sent home for a week to recuperate.

Since I returned to my station after knee surgery, things were not going well. The guy who drove while I was off for six months liked the extra money he was paid and hated giving it up. His answer to that was to be mean and nasty to me, hoping I would put in a move and he would get my position.

One of the other guys tried incessantly to get in my pants. He was unhappy that I was telling him no. He followed me down to the basement where I was loading the washer with my dirty laundry, pinning me between him and the washer. Another guy walked in and caught him putting the moves on me. Immediately, he announced over the loudspeaker how we were rocking with the spin cycle. It thoroughly embarrassed us both. The next time he cornered me, I wasn't very nice about saying no. He stopped bothering me for sex but sought revenge.

The young, hulking, near sub, (gone regular recently) whose opinion of women was borderline, was influenced by these two. He joined them to make a trio of guys who constantly picked on me. I couldn't mop the floor right. When I cooked, hair mysteriously appeared in the food, supposedly from my head, even though I wore my hair in a ponytail. If I had friends over, they griped because the males would use my bathroom, even though the other woman didn't care. It was none of their business who came into the women's locker room.

Some of the other guys in the house would take my side and pick on them for the dumb stuff they did. One guy picked on people randomly. War broke out with everybody picking on each other. The loudspeakers constantly blared with insults to somebody.

The female firefighter stationed with us was a young, beautiful, African-American woman who had been to medic school. Among her many talents, she was a former cheerleader for the professional football team in the city. She and her classmate were as nice as they could be in stark contrast to the three who were pains in my ass.

When it was my turn to cook, the three troublemakers conspired to give me pennies, nickels, dimes, and a few quarters to pay their six dollar meal money. I stopped at the bank to cash in the change, drove around the district to some peculiar addresses to practice getting there, took my time at several stores shopping for sales, and served lunch late. The trio were mad, but the other guys liked my ingenious payback. The three inspected the grocery receipts proving I paid by credit card. This confirmed I didn't have to go to the bank to cash the change which inflamed their griping.

Retiring to the locker room where the other woman was taking a break, she took the phone off the hook preventing everyone from shouting insults over the loud-speaker. I was upset, feeling that "everyone hates me".

She gave me a hug and said, "A few jerks can make it seem that way, but it just isn't true. Almost everybody thinks you are great!"

We sat in the locker room together, reading books until it was time for me to cook dinner and serve up hair in the spaghetti sauce.

We didn't set up our aerial on runs often because most fires are in small houses. One month though, we had to set it up three different times. We had a fire at a gas station with the potential for a big explosion. Next, we had a fire in a business in a strip mall. Finally, we had an apartment fire. It was a few minutes from the firehouse, but the flames were already shooting out the second story of an end apartment when we arrived.

I positioned the aerial, forward facing to the fire, allowing me to raise the ladder in line with the chassis. This prevents the truck from tipping over if a couple of people are hanging onto the end of the ladder. A woman ran over, screaming that people were trapped on a balcony a few doors down. Three guys grabbed the 35-foot ladder and rushed off to help the victims, leaving me to raise the aerial alone. I got the stick (the aerial ladder) up, placing it inches above the balcony. Resting it directly on the balcony could be detrimental to both objects.

The decision was made to battle the fire from the interior since it appeared to be confined to the one apartment. An outward attack could push the flames to adjoining apartments. An interior attack would keep the flames confined and move them exteriorly. I was no longer needed to stay with the aerial since we wouldn't be throwing big water, but as it should be, we were prepared if the fire progressed.

The chief assigned me, along with three others, to search all of the apartments and clear out the building. Two of us went up and two went down. We knocked once and then busted in the door. There were approximately 12 apartments per floor. The fire end of the complex had smoky apartments that were slow to search. As we moved away from the fire, we were able to speed up because it was easy to see the places had been vacated. Some doors were unlocked, and we didn't have to break them down. The ones that were locked were easy to get into, one swift motion with my hip against the door, popped it open.

In the second to last apartment, after we knocked, received no answer, and broke in the door, we discovered a resident still in there. Since it wasn't very smoky, he decided he didn't need to evacuate and miss his TV program. With his calm manner, we could have used him on the department. We escorted him out and told him he could watch the fire on the news later, but for now, for his own safety, he needed to get out. We reported all clear on the search.

Pretty soon, we were cleaning up and going home. After every fire, we review the defense. My partner told everybody how I was Bertha Butt, knocking down all the doors with one swing. The guys who put up the 35-foot ladder complained about their backs and how heavy that thing is. We were all laughing. There is nothing like a big fire to bring peace and unity in the firehouse. The next day, everything was back to normal, a little sniping here and a lot of griping there. Some houses never find permanent peace.

Another fire, shortly after that, started in a kitchen and was contained there. The flames ate away part of the ceiling and somebody needed to check the attic. It is always tough to get in the attic. Most of the time, the SCBA tank has to be removed and pushed through the scuttle hole, the firefighter following, then putting the pack back on. The attic clears out last and is hotter than hell. I, as usual, volunteered to go up and check it out. The lieutenant from another company heartily nodded his approval, better me than him.

It was smoky. Another big guy followed me up but didn't have his mask on. He went back down quickly with watery eyes and a snotty nose leaving me by myself. Careful not to put a boot through the ceiling, I crawled over the rafters, checking for fire extension. I found a gable vent and saw the smoke whisking out. Pretty soon it would all be clear. I went down the ladder reporting the good news to the officer. On the way out, the hulking backstepper gave me a slap on the back, saying he was proud of me for checking out the attic. He bragged about me back at the firehouse.

Weeks later, our Hulk came to work day after day in terrible moods. He constantly griped at me and got the trio going. As he watched a country music station, he called one of the women singers a "bitch" because she had too much of a "pop rock" sound.

When a fire came in, he said, "Follow me. I'll show you how it's done."

I said, "Go ahead big boy. Let's see if you can teach me anything at all."

The captain was on vacation. The acting officer, part of the trio, and the other backstepper went on the roof. The

Hulk and I were assigned search and rescue. We were a second arriving company. The fire was on the second floor, with a line of fire fighters waiting to go up a narrow stairway in the back of the house. Since the Hulk was anxious to show me how to be a "fireman", he was fidgeting around trying to squeeze by some of the other people in line. Nobody let him through. He stormed out ordering me to follow him. We got a ladder and put it up to the roof over the porch in front of the house. We walked on the roof to a window and opened it. He hesitated, feeling the heat and seeing the flames licking around the back of the room. The fire was being put out though, by the engine company crowded at the top of the stairs.

Going back down the ladder, he said, "We'll get a line and bring it up here."

He raced off to get a hose and I bee-lined back in the house, up the stair well that was passable now. I helped the engine crew with search and rescue. The guy on the nozzle was griping that my team wasn't there, when I appeared. He asked me where my partner was and I said he wouldn't come in. A member of his team and I searched the floor that was still as hot as when the Hulk tried to enter through the window. We reported all clear and went outside for a break having doused the fire. There was the Hulk, still outside and pissed because I hadn't followed him to get a hose line. I told him, if he needed a line, then he should put in a transfer to an engine because truckies don't use hose lines; truckies use axes and pike poles. He called me a bitch. I called him a dumb, fucking, near sub who is nothing but a dickhead.

Some firefighters witnessed the exchange and said, "Can you feel the love?" They left laughing.

Back at the firehouse, the Hulk caught me in the kitchen and screamed at me for leaving him, lecturing me on how we are supposed to stay together. I screamed at him that he showed me nothing, and I didn't want to be his partner if he was afraid to go in. I left the kitchen with him yelling at my back. All kinds of comments went flying, pro and con, over the loudspeaker. It stopped only because it was time for us to go home.

Our next fire was in an old, abandoned house. We arrived as a second in company. The first crew attempted to enter through the front of the house, but backed out, reporting the floor unstable. Everyone on the firegrounds was warned by the safety officer not to enter. My two, macho partners, parts of the trio, entered anyway, dragging a hand line that was abandoned outside the door. The skinny guy was first, the Hulk next, and I brought up the rear, pulling the hose as we entered. Halfway into the room, as we walked, not crawled, the floor gave way and our lead guy disappeared through the hole. The Hulk and I both dropped to the floor, looking through the hole, yelling for him. We saw he had fallen approximately eight feet onto a concrete floor in the basement.

The Hulk screamed, “Are you okay?”

The skinny guy yelled back, “I hurt my ankle!”

All of this was hard to understand because we were yelling through our masks. The fire was in the rear of the basement. A crew who was down there, and had put out most of the fire, helped him hobble out. The Hulk and I crawled out with the floor buckling under our knees. We found our skinny guy sitting on the back step of an engine drinking a sport drink. We helped him get his gear off and tried to talk him into going to the hospital. He declined saying his ankle felt better.

Winter driving is challenging in the big rigs. They slip-slide easily if the streets are wet or icy. I was driving my crew in the new aerial on slick streets. The engine was delayed getting out so we were in the lead. I needed to turn just a few streets after we left the firehouse so I did not go to warp speed. There was a light dusting of snow on the street when I put on the brakes and started my turn. The truck went up on two wheels. I immediately let off the brakes, straightened the wheels, and the aerial sat back down on all her tires. I drove to the next street, making my turn very slowly. It was a false alarm, and we went back home. The way the trio talked, you’d think I had rolled it a couple of times. If some other guy had done the same thing, they would have patted him on the back, congratulating him on the great recovery he made!

We were dispatched to a tricky location. Since I studied a map every day, I knew where it was for a change. The engine was in the lead and passed the turn. I turned in the right direction arriving well ahead of the engine. It was a false alarm, so it didn't make any difference, but the engineer still felt badly. He was moping around the kitchen after we got back. Being part of the trio, my two backsteppers sympathized with him, patting him on the back saying it could happen to anybody.

"It was a shame you heard the address wrong. Everybody on her truck has to pay attention all the time since she is liable to go wrong any time."

My roofer came through and told the trio, "That's what you get for throwing stones, because sooner or later, everybody makes a mistake and then you look like asses."

One of the guys who often took my side was from Boston with a heavy accent. He had a heart of gold. At Christmas, he brought in two little bikes and assembled them. Since he was single, everybody wanted to know if he had some kids we didn't know about. He said he wanted to share his good fortune with others, so he was donating the bikes to Toys for Tots.

Every year we get comprehensive physicals to try to cut down on the number of heart attacks on fires. Once in a while, a firefighter flunks the 12 lead ECG (spelled ECG for electro cardio gram but commonly pronounced EKG) on the treadmill. When the doctor sees an abnormality, the firefighter must get his heart fixed before returning to work. Even though it has saved a few lives, everybody hates the possibility of being barred from duty. Our blood is tested, hearing checked, urine analyzed, resting and stress ECGs are given, and the dreaded body fat percentage test is taken. When the results came back, the trio called me "quarter fat". When the Boston backstepper and the other woman firefighter said they were 25 percent fat, the trio shut up. (The woman backstepper had to be fibbing in order to help me out because she is slim.)

Responding to interstate runs can be as hazardous as running into burning buildings. Some people are busy rubber-necking at accidents and don't see us as we execute our job. Dispatched to a highway location we pulled over on the berm at the scene of a car wreck. Everybody denied injury and needed no help. Since no one complained of any problems or pain, my officer called it a property damage accident and disregarded the ambulance. Our truck was on the run with the engine because entrapment was reported.

Preparing to leave, I climbed into the truck. Our Boston guy was standing by my door with his back to the traffic. A car drove by, clipping his back with its side mirror, nearly knocking him over. The car almost took my door off. He screamed at the people in the car as I jumped out of the truck to get a license number. We thought the car was going to stop and pull over but it sped off with the two of us in foot pursuit, about as likely to catch the car as a kid to catch a bird. He was cursing as he ran, worse than anything I ever said, trying to get the plate number. Neither of us could read it. Since he rarely ran, I knew he was okay. He was aching, but refused to go to the hospital. I asked him if he could teach me some lessons from the book he must have read, "The Nastiest Possible Curse Words on Earth".

The captain was a nice man. He was newly promoted and served most of his years as a lieutenant in the office. He didn't know what to make of all the infighting. He refereed as best he could. He believed, at first, if he brought the combatants together to hammer out the beginning of a solution, things would improve. For the first session, he brought the skinny backstepper and me together. It started calmly enough, but escalated to a screaming match with both of us accusing each other of dastardly deeds. The skinny guy swore I was kicked out of my last firehouse. I asked him the number of firehouses he'd been regular at, which I thought to be about nine compared to my three. So, who had trouble getting along? In my entire life, I never pulled a gun on anyone. He had at one of his rentals. Who got handcuffed by the police? Not me! Who put the aerial on two wheels? Not him! After a few minutes of this and the loudspeaker starting up, the captain sent us to our separate corners. The captain didn't try that method again.

For Christmas, the captain gave every body little tree ornaments, a dog wearing a fire helmet holding a hose. We quit fighting long enough to exchange holiday greetings, but I had enough and started watching the vacancy list. Informing the captain of my intention to leave, he asked me not to go but I felt it would be better for the house (and me) if I found a new home. Maybe some of the fighting would stop if I was gone.

Part Six

Down the Homestretch

Chapter 31

Not long after I decided to get the hell out of that station, my old chief, one of my favorite people in the world, stopped by the firehouse to say hello. Even though he didn't live far from there, it was unusual for him to drop in. It was a nice, surprise visit. He's a very popular chief, and it caused a few murmurs among the trio. Showing him the new aerial, he took my picture in front of it which is on the inside flap of the back cover.

He came by to ask me a favor. An engineer would retire in a month from one of his companies, and he wanted me to take his place. I was ecstatic, the chief asked me to come to his district. I told him since my knee surgery, truck work was tough on me and I wouldn't mind the easy life of an engine engineer, and it would be great to see him once in a while again. His station was closer to my home, making the commute shorter. I would leave the trio behind, but I would also leave my roofer, Boston buddy, women friends, and a nice captain. The lieutenant on the engine was retiring soon. I would miss him either way. When a chief asks you a favor, it is hard to say no. I had an excellent excuse for putting in my move.

A few months later I was on my way to the south side. The skinny guy of the trio didn't rejoice very long. My vacancy was filled by a person from another company with a lot more years on than him. His plan to run me off and get my spot failed. I had the last laugh for the moment. He is a captain now and laughs all the way to the bank.

My ride to work was 15 minutes shorter each way. It isn't really that important since I drive to work just nine times a month or, including vacation, about 100 days a year. With the rising price of gas, fewer miles are good.

Since my knees are beat up, I looked forward to being an engineer on an engine again, standing outside pumping, not having to go in and do arduous work. Engine engineers have cushy jobs. Since I know the district, I wouldn't have to worry about waking up and racking my brain for where in the hell I was going. I was sure I would sleep better, but there were a few things I hadn't counted on. The noise from trains, automobiles, street sweepers, and garbage trucks keep me awake at night. This firehouse sits so close to the street that you can't park the engine on the approach and shut the overhead door. The bedrooms are in the front next to the road where every thump-thump car radio can be heard. It is squeezed between railroad tracks and the back of a grocery store where the dumpsters are located. Every night at midnight the train rattles the beds as it roars past. At 0500 every Wednesday morning, the garbage truck kathuds the dumpsters as they are emptied. On Thursdays, 0200, street sweepers clean the parking lot of the store. I hear them going back and forth as I lay in bed. I believe this is the noisiest house at night on the department. I wonder how the people in the houses across the street cope with the cacophony. Maybe they block it out because they deal with it every day or maybe I am a light sleeper.

The four men at my new station were a perplexing set of personalities. One was an ex-Olympian, rock solid, compact and a health nut. For his day off job, he was a partner in a gym with the local, football, weight trainer. We read the paper every morning and ate breakfast together. He would get a mixing bowl, filling it with a box of cereal and a quart of skim milk, and then eat the whole thing. We sat outside together in lawn chairs, talking and tanning. He called me Princess. His daughters are Precious and Cutie. He was never in a hurry, always the last one to board the engine before takeoff.

His philosophy was, "It was burning when they called; it will be burning when we get there."

This flustered the lieutenant and me. We wanted to hurry. Why delay? Fire expands exponentially. It only makes things worse. Any delay only causes mega damage and more work. No amount of prodding or begging moved him along.

The other backstepper didn't socialize. He went to the front room where he stayed day and night, smoking his lungs out. He came out to cook dinner, which was the only meal we ate together. Lunch was whatever leftovers you could find from the other two shifts. There were always plenty of leftovers. If I had to go into the front room for something, he barked at me. I left him alone. He was not a bad guy, just a little grumpy. He told great stories about anyone we were discussing. (If you haven't noticed, we gossip a lot.)

The lieutenant played melancholy bagpipe tunes. He is a member of the fire department bagpipers who play at funerals and memorial services. I get chills when he plays "Amazing Grace". The station is in a nice part of town, with better educated citizens so it doesn't get very many fires, although it has had some doozies. Before my arrival, the propane gas company, a block away, erupted with exploding tanks flying everywhere. The lieutenant got a commendation for his cool, effective leadership at that memorable fire.

The populace is aging, so we get a lot of chest pains and difficulty breathing calls. We also have what is called a brittle diabetic. He is one of our frequent flyers. We respond weekly to his house. Our fifth person roved a lot. He is a red headed, Native American, also a champion weightlifter. We didn't see him much.

Since the inception of modified response, (one engine and a truck are dispatched for unsubstantiated alarm runs) this company averages a little over four runs a day. Of all the companies in Indianapolis our average puts us in the middle for runs dispatched. We get few runs after midnight, frequently sleeping through the night.

At budget time, talk goes around that the administration is considering closing this firehouse. Several houses have been closed since I came on. A couple of chief positions were created and the budget needs balancing. They cut the worker bees to do it. Most of the guys in the office would climb over their grandmothers' backs to get to the top, cutting positions to please the mayor, their appointer.

The firehouse is important to the area. We've had several, small fires at the college next door that could have been disastrous but our two minute response time averted the nightmare. The senior community around the corner calls us almost every day for assistance. Sometimes it's just, "I've fallen and can't get up", but frequently it is serious medical problems that need quick responses. Ambulance arrival is delayed due to their distance from the scene.

Since most of our runs are medical and ambulances take a long time to get there, I decided to, once again, apply for paramedic training. When I asked the guys at my firehouse what they thought, they were enthusiastic about my plan. They worried there might come a day when every house and shift would be required to have one paramedic on duty. They said I was the smartest of all of them and they could stop worrying that they might be sent. (I think the latter was their motivation and the rest was flattery to get me to go.)

Chapter 32

Predictably, the process changed since I last applied. The new chief employed reading and math aptitude tests to measure the learning ability of applicants. The chief decided this was the best method because any weaknesses in EMT skills would be strengthened in paramedic school. I was finally in because I am good at learning from books. I established my real estate business, which is now a multimillion dollar operation, by reading books. I did well in college which consists mostly of devouring books.

Around this time, I met my soon-to-be-husband. We saw each other on runs, but had never spoken more than a hello. He is a police officer and a total babe in his uniform. He visited the firehouse once in a while with a cop buddy. I drooled every time I looked at him. He is a shy, quiet person. I was my outgoing self, telling dirty jokes and raucous stories. One time, he got up in the middle of my narrative to leave. I worried that I said something to offend him. When he came back he explained he left because he got a run. Police radios have constant chatter.

I use the parking lot next to the firehouse to exercise, speed walking around the grocery store with my headphones blasting in my ears and the portable radio in my hand. I love working out to music. "Mony, Mony" makes me walk faster. One morning, the babe pulled his patrol car beside me and rolled down his window. In order to hear him, I took off my headphones. He asked me for a date. This handsome man said if I looked good in bunker pants, he could only imagine how I look dressed up. Since I am much older than he is, I turned him down.

He said in German, "Ich habe aeltere Frauen gern." (I like older women.)

I answered, "Du kannst ja Deutsch sprechen. Wo hast Du das gelernt?" (You can speak German. Where did you learn it?)

Not expecting an answer and embarrassed that I understood what he said, he was speechless. Continuing in German, I asked him if he had been to Germany. He said he learned to speak it as a kid from his next door neighbor who emigrated from Germany after the war. I told him about my three year stint stationed there and perfecting my German while I was in the army. He was jealous because he wanted to go there but got stationed in North Dakota for his Air Force duty, a place that wasn't mentioned on his dream sheet. I was in the process of severing a relationship with my boyfriend, so his timing was great. We moved in together three months later and were married three months after that. Some people thought I was pregnant (my parents) since it was so sudden after 14 years of being divorced. I had declared I would never get married again. Gladly, I ate those words. We had a nice, little ceremony by the lake where we lived on a warm, beautiful, October afternoon.

My concerns about paramedic school were, being a newly wed, I would have to take time away from my husband to read and study. Secondly, I would have to leave the companies for 16 months while in school. I wasn't sure I wanted to do that. Our education chief encouraged me to attend. Most of the applicants were young and inexperienced. They wanted some senior people in the mix. I was already committed to an ocean cruise during the time we were scheduled for class. With a promise from the chief that I would be allowed to go on vacation (I had already paid for the cruise), and a push from my husband, I signed the papers committing me to the process.

It was a big class of about 20, with students from other departments and ambulance services. We lost four in the first few months during anatomy and physiology. Anatomy was one of my courses in college, but physiology was very foreign. I studied until my eyes were bloodshot.

My husband noticed this and took me to an optometrist for glasses. He said he was tired of seeing me squint to read. The squint furrow between my eyebrows was growing larger on a daily basis, and I got headaches. Although I hated to admit it, I was relieved to get glasses. My eyesight had always been top notch but everything seems to slowly go haywire as we get older.

While I was away at school, the weightlifter at my firehouse found a lump in his neck. The doctor took a biopsy and pronounced it cancerous. During a short break at the hospital, I called the firehouse to see how things were going with everyone. They told me how he was whisked away to a hospital for chemotherapy. I burst into the instructors' offices, breathing heavily. As I tried to tell his story, the tears started rolling down my cheeks. They sat me down, giving me a minute to calm myself. They let me telephone him, my friend who called me Princess. He did not want to see anyone because the chemo was making him vomit all the time. He was proud of his former physique. Now he felt emaciated and embarrassed. I didn't care about the condition of his body. I wanted to see him but respected his request. Immunocompromise is a side effect of chemo making it dangerous to patients if visitors bring in germs. Putting aside my selfish wants to see him was the best thing I could do for him. The fewer people he saw, the less likely he would be infected by a virulent bug.

I went on my cruise while he was getting radiation treatments. It was great to get a break from school. I took some books to study but barely cracked them, having no will power and too much fun to study. When I returned, I was behind but made it up with extra effort. The good news that my weight lifter was on the way to recovery, relieved me. We were closing in on Christmas break, and the doctors predicted he would be back at work by then.

The two week school recess allowed us recovery time at the firehouse with friends. My weight lifter was pronounced cured by the doctors. When he told me he was getting a touch of the flu, I became instantly concerned and told him to go back to the doctor immediately. In paramedic school, we studied the body's reactions to various diseases. A person may have a flu-like reaction to cancer or it could just be the flu. For my peace of mind and as a precaution, I showed him in the book what I learned could be a dangerous symptom. He fought through feeling sick, not missing a day of work or going to see the doctor. He was gaining back his weight, looking good. I quit worrying.

The new semester of school was exciting. We started ambulance rotations, practicing our newly learned paramedic skills. We were let loose on the public under the guidance of seasoned paramedics to: start IV's, give medications, analyze heart rhythms, administer breathing treatments to asthmatics, and intubate unconscious victims if needed. In school, we practiced all that stuff on plastic dummies. On the ambulances, we advanced to real people. In the hospital, when I had trouble with IV's, the nurse took me in a room to practice sticking the needle in her arm. With my sweat dripping on her arm as she coolly accepted the sting of the needle, I improved.

We had interesting ambulance runs. A middle aged woman experiencing chest pains went to a clinic for tests and began to have severe pain. We were called to transport her to the hospital. We administered oxygen, nitroglycerin, and aspirin. Her blood pressure dropped dramatically. To alleviate that, we elevated her legs and gave her extra IV fluids. In the emergency room, the doctor wanted to know why she had a precipitous drop in blood pressure. Had she taken a sex enhancing drug, given mainly to men for erectile dysfunction but women can take it too? I already checked and knew she hadn't. The doctor wanted my opinion. Did I think it was a heart attack? I gave a long, noncommittal, indirect explanation, sort of a politician's answer, that it could be a hundred different things. He made me commit to yes or no. Hedging, I said yes, even though the odds were against me. He finally let me go. I didn't find out if I was right. When I saw the doctor, weeks later, he didn't remember the patient's outcome.

My weight lifter went for his annual wellness physical. During a routine chest x-ray, cancer was found in his lungs. It had been there all along. He had six months to live. Receiving the news at the hospital again, I ran to a patient bathroom to sob. He deteriorated rapidly. I would never see him again. Cancer takes too many firefighters. Even though we are generally more fit and have a slightly lower incidence of heart attacks, we are ten times more likely to die from cancer than the average civilian.

On the ambulance we were dispatched for a grade school age boy who fell down the steps at home. He had only a small lump on his head. What concerned my trainer was our difficulty in rousing him. His mom claimed he wasn't acting normally. In case of spinal injury, we had to stabilize his head and neck which meant tying him down to immobilize his body. He woke up during the process and not understanding what we were doing, started screaming and fighting. Kids don't like to be held down. It scares them. They panic, and it is almost impossible to get them to calm down. We wheeled him to the ambulance as he screamed. His mom cried too. We put Mom up front with the EMT driver, buckling her in for the ride.

As we took off, I wanted to decrease this boy's anxiety. Reassuring him that his mother was with us and putting her hand in his, I got a teddy bear (the police, fire, and ambulance services carry stuffed animals) and held it close to his face. My trainer obviously didn't agree with my method, because he grabbed the bear out of my hands and threw it against the back wall of the ambulance, which made the kid scream even louder. Through clenched teeth, he told me to get a blood pressure and other vitals and to stop playing with teddy bears. My protestations that keeping him from crying would lower his blood pressure, which would keep his intracranial pressure from rising, which would be good if he had a head injury, were met with stern glares. I did as ordered.

Not wanting to start an IV on this kid, I appeared busy, so my trainer had to start one. I sure as hell didn't want to add to this boy's misery by attempting an IV needle stick in the back of a moving ambulance, and miss, causing him to be stuck a second time. My trainer's manners may have needed some fine tuning, but I trusted his ability to start IV's. When any patient was difficult to get an IV started, he was asked to help because he was known for his ability. After he got the IV started with the first needle stick, we had to tie the boy's arm down to prevent him from accidentally ripping out the IV. That was the most terrible ride I ever had to the hospital. Giving CPR to dead people in the back of ambulances does not compare. They don't scream, cry and kick, melting hearts with their suffering.

That ride conjured memories I suppressed when I was in the academy going through EMT school. We were required to spend 12 hours in a hospital emergency room. One of the patients who came through was a young girl, bitten in the face by a big dog at her parents' friend's house. Her cute, little cheek gaped from the wounds and a chunk of her ear lobe was missing. Her dad was sent back to the house to look for the chunk of ear. Her mom waited in the room, just around the corner. As soon as the doctor began to examine her face, she started screaming. He called for a plastic surgeon who could minimize the scars. The specialist reported to the crying kid and ordered her to be put on a papoose board. When asked to get it, I had to admit I didn't know what it was. A nurse ran off and came back with it.

I understood how it got its name. It looks like an oversized version of what Native Americans carry their babies in. Putting her in it made her scream louder. She jerked her head back and forth as we tried to hold it still. The doctor ordered a kiddy cocktail, a sedative that quieted her. We thanked our lucky stars for the peace.

Now that it was quiet, the nurse heard me sniffling and asked if I was okay. I squeaked out a yes, dabbed at my tears and watched as she was deep cleaned, then stitched with a curved needle. The chunk of earlobe was not recovered forcing the surgeon to stitch together her ratty ear. She looked miraculously better when he was done. I wondered if that piece of earlobe ended up in the dog's stomach; if the dog was destroyed; and if the people remained friends. Everybody was a loser in that situation. No wonder I buried that one.

Sometimes we are called to help people who don't want help and refuse to be transported to the hospital. Two instances of this occurred with my paramedic trainer and me. Called by a mother, we arrived to find a 50-year-old man on the floor of his trailer. He was sick and unable to get up, lying in his own feces. Attempting to sit him up made him feel worse like he was going to vomit. His putrid smelling body was gagging everybody. He begged us to let him lay there. He wanted to sign the paper which releases us from liability so we would go away and leave him alone. He knew the process because he had been through this before.

My trainer said, "I won't let you sign the papers. I will not leave you here to die. You can sue me in the morning from your hospital bed."

He called for help to get the patient loaded. As we waited outside (we needed a few breaths of fresh air) for a truck crew to arrive, I questioned my trainer about ethics in taking this person against his will.

He said, "I don't care about ethics. I will not leave a person to die without intervention. The patient can sue me and the hospital can fire me but I can't let somebody in his shape sign an SOR."

Later that week, the police requested our assistance at an automobile accident because a person complained of difficulty breathing. We arrived and found a beautiful woman, in a red convertible, answering us in a hoarse whisper. We did our best to convince her to go to the hospital, but she refused saying she often has this difficulty when she gets upset. She was a patient of a specialist who couldn't figure it out but deemed it non-life threatening. We spent an inordinate amount of time on the scene, intimidating, foreboding that she might die. Still, she refused. She accepted a few whiffs of a breathing treatment. When my trainer said she would have to go to the hospital now that we were treating her, she threw the oxygen mask out the window and rolled up her convertible top with a switch inside the car. The police officer crossed his arms refusing to place her under arrest so we could treat and transport her.

I got my trainer by the arm and attempted to walk him to the ambulance. The police left. My trainer shook me off and grabbed a knife out of his pocket, threatening to cut the top of her car and pull her out. She called her fiancé who arrived a few minutes later with the police again. Her fiancé said he would take her directly to the hospital. Entrusting her to the care of her boyfriend, we finally left. My trainer and I argued about the whole situation. He asked me not to come back. I finished school with a different ambulance crew. Eventually he lost his job over his actions, even though he had been a paramedic over 20 years with them. There is a fine line between good patient care and overzealous behavior.

My new trainer was a medic with the police SWAT team and highly respected, an excellent instructor. I felt lucky to be on his ambulance except for one drawback; my new hours were from 1500 to 2300. Since my husband goes to work at 0530, or O dark hundred as he calls it, to 1400, I would not get to see him much. I was close to completing my requirements. I would have to bite the bullet, so to speak, for a short time.

On most of the runs with my new trainer, I was the team leader. Being a book smart, almost-paramedic, I got a boost the day I taught my veteran trainer a little something. We had a chest pain patient taking a prescribed blood thinner. After asking all the requisite questions, I gave him aspirin along with nitroglycerine. My trainer said not to give him any aspirin since he was already on blood thinners. I explained "sticky balls" to him. Red blood cells clump in people's heart veins forming clogs (sticky balls) which will eventually stop blood flow to parts of the heart muscle causing a heart attack. Aspirin inhibits red blood cell stickiness so less clumping occurs. Blood thinners act in a different way, making aspirin a necessity. The doctor confirmed it. My instructor thanked me for teaching him something new. It was refreshing that he wasn't afraid or ashamed to admit he didn't know it all.

One night, he was called to a briefing for a SWAT team action. I was goose bumpy with excitement as we reported to a police center for the diagram and plan of action. Several of the officers said "Hi", because I knew them from prior runs on the street. The intricate planning was different for me. As firefighters, we run in like gangbusters and sometimes talk about it afterwards.

With the plan set, we ditched the ambulance and were chauffeured to the location in the back of an unmarked police car. It was an old jalopy, driven by a greasy, long-haired, undercover, narcotics cop. The operation was a drug bust and SWAT would sweep the house making arrests along with other agents. The officers said they had made several buys and watched the house for a long time. This culminated a long effort that the neighbors were anticipating. The residents were sick of the drug dealers poisoning their kids. I watched it all from the back seat, a few car lengths away from the house. Months later, at a ceremony where my husband was receiving an award, many of those officers were honored for this bust.

The next SWAT team action was in progress when I reported to the ambulance. My trainer's partner was waiting for me to hop in. We sped off to a site with a sea of black and white police cars, a huge van, and news media crowding around. We got through the maze, squeezed to the front and entered the van, a rolling police lab. We witnessed the events on monitors or with binoculars. The airways buzzed with chatter as cops maneuvered toward the felon hiding in a house. They progressed slowly, coming to a standstill. During the inaction, the person operating the van showed me the fancy equipment stored all around, explaining each item's purpose. It was hours before a final move was made to storm the house. My trainer ran in with the SWAT team. Since no one got shot or injured, his services weren't needed.

All that waiting around made me hungry. As soon as my trainer was released, we headed straight for a restaurant and then we chugged some coffee at their favorite bistro. They also opted to make a contribution to the state and bought lottery tickets.

When I think about it, I am immensely glad I became a firefighter and not a police officer. Everybody is happy to see us arrive. When we put out a fire, we are buddy slapped on the back and congratulated, feeling a sense of elation, and most of the time, no investigations. If they shoot somebody, they are given the third degree by Internal Affairs. We wait around, comfortable in our firehouses, for a run. The police wait around in ditches and behind trees or in their cramped police cars. The president of the police union reputed our firehouses as palaces, arguing with the mayor that firefighters have it cushy compared to police. I agree with him. Where else will you get paid to talk on the phone, watch TV, do your laundry, cook and eat meals, read or even write a book? On the other hand, more firefighters get killed in the line of duty than police officers.

The requirements to graduate from paramedic school include multiple exams that last for days. After a written test, that makes no sense and is unbelievably confusing, there are a series of practical exams, forcing students to prove they are capable of performing each required paramedic skill. It's nerve racking because we are placed in a room alone with a test proctor who watches every move we make, questioning the would-be-paramedic if she seems uncertain.

One of our classmates wasn't allowed to test because he didn't get his rides on the ambulance finished in time. He also had some emergency room hours to complete. He was way behind. I thought he might get in trouble for dereliction of duty. He was paid to get everything done but didn't. Would it be considered the same as missing a firehouse duty day or blowing a run? We all waited, placing bets, to see if thunder rolled on high and lightning would strike him down.

I missed criteria in two segments of the practical test. Missing four segments, fails the student and he must test another day, no sooner than 30 days and no later than one year. I retested the segments the next day and passed. Several classmates retested and failed. They were scheduled to retest the following month.

Graduation ceremonies were held the week after primary testing concluded. Everybody participated because it was anticipated that all of my classmates would eventually pass the tests, which they did, except for the one who was far behind. He completed the requirements but didn't pass the test in the required time frame and chose not to go through the course again. Since he completed the practicals, lightning didn't strike him. It made me wish I hadn't busted my butt to get done on time. It would have been much less stressful to take a little more time to complete the requirements.

I graduated number two behind a young, beautiful, female firefighter. She got as skinny as a rail, and I got as plump as a little piggy. She starved her way through school. I ate my way through. She was asked to give a speech about 30 minutes before we walked onto the floor. It made me glad to be second. Her extemporaneous speech was eloquent. One of our firefighter/paramedics died in a training accident two years before. The award for class valedictorian was being given in his honor. She knew him and talked about his fervor for the job. She said she was inspired by him to become a paramedic telling a short anecdotal story which made everyone laugh. She discussed our hard work and dedicated it to his memory. His father was there, and with tears shining in his eyes, he bestowed the plaque. We clapped enthusiastically.

The public safety director, a former police officer, spoke next. He forgot his audience and spoke about the police department barely mentioning the fire department or ambulance service. We clapped politely. What an idiot! But his irrelevant speech couldn't spoil the moment. We were all returning to the companies. In a few weeks, we would start practicing our paramedicine. I hoped the 16 months of intense study would be worth it. In a year, I would pass judgment. The Chief of Personnel, the captain from my old station, was there. She posed with me for a picture, the first female chief on IFD hugging the first female engineer, who was belatedly and ecstatically graduating from paramedic school.

Chapter 33

I sobered up at my station when the drug box was delivered to me because it was full of expired medicines and not much else. There was no list of what I should have in it. When I called the office for help, nobody returned my phone calls. Finally, I called one of the paramedics who is a longtime friend and she talked me through my box by looking in her box. I ordered a long list of supplies, which didn't come in the next three days. I reordered them and they still didn't come. The lieutenant in the office called and wanted to know why I was not marking my apparatus ALS. When I told him the medicines weren't delivered, he called the supply desk for me. All three orders were delivered the next duty day. Swimming in medical supplies, I wondered where I was going to put all this stuff. The house captain ordered a locker for me, and we shelved it all. It looked like a mini pharmacy.

After I updated all the drugs in my tech box, I spent 60 minutes with my crew, summarizing my 16 months of schooling and enlisted their help, which they were happy to give. After my little speech, we laughed that I could review in an hour what took me so long to learn. Calling Control, I marked, for the first time, my engine as an advanced life support apparatus. It was exhilarating, yet daunting. I would now direct the actions of the crew on medical runs.

While I was away at school, my old officer put in a move. I was breaking in my ALS skills and a new officer. He is an ex-Navy seal, solid as a rock, but soft spoken, a drinker and a womanizer. Being a newlywed, I was not interested in him. Being an astute man, he never bothered me. Cranky Butt, our backstepper and station cook, didn't like the new boss. His appearances outside of the watch room became even less frequent. Sometimes he refused to cook, which made the two of them bump heads. It wasn't so maddening that he wouldn't cook but that he didn't say he wasn't cooking, which left us scrambling for meals.

With our ex-Olympian weightlifter struggling for each breath at home, our fifth person was at the station every day in his place. Our Native American firefighter practiced his heritage. He brought deer skins to sew into moccasins or drum heads, patiently working on an interesting project in front of the TV in the kitchen or in the apparatus bay. I went to some of his pow-wows to watch him sing and dance in the native clothes he made at the firehouse. In between projects, he ran to the washer and dryer in the basement to throw in more clothes that he brought from home to wash. It could be difficult for the rest of us to wash our dirty uniforms because we had to squeeze our one load of laundry in between his multiple loads. He sought advice about real estate investments and we discussed the pros and cons of being a landlord. He took the plunge, renting his old place and buying a bigger home for his beautiful Hispanic wife.

A middle of the night alarm rang for a person not breathing. We headed straight there as I reviewed in my head the procedures and medicines I might administer. Procedures are always changing, but at the time, the first action was to hook the patient to a monitor if they are not breathing and you can't feel the pulse. The monitor showed a rhythm that needed shocked. I told everybody to stand back, hitting the shock button but nothing happened. I hit the shock button again and again, no shock, no patient jumping from the jolt. Feeling the panic rise, I took some breaths to stay calm and ordered CPR started while I got out my equipment to put a breathing tube down his throat.

The ambulance arrived with a paramedic who started an IV for medicine administration. Our patient went back into a shockable rhythm, but we hadn't switched machines and couldn't give him a jolt. The paramedic punched him square in the chest (precordial thump). He converted to a regular rhythm which confounded us. Precordial thumps rarely work. This paramedic was good. We quickly switched monitors in case he needed another shock. The pure oxygen we administered was obviously helping because his heart kept ticking away and we could feel his pulse. Sometimes the monitor shows a beat but a pulse can't be felt. This is called pulseless electrical activity and is not good. The medical advice is: treat the patient and not the monitor. A beat on the monitor without a pulse requires special heart medicines. Always check monitor activity against patient vitals. Our patient got the correct meds. We loaded him on the cot transporting him to the nearest hospital with lights and sirens going, sorry for the people we woke as we passed their houses at 0200.

Returning from the hospital, at 0300 I called and got the sleepy but helpful emergency duty officer out of bed to tell her my equipment failed. At 0400 she was at my station with a new heart zapper. We went through the procedures to ensure I operated it correctly. Being fresh out of paramedic school, I was familiar and comfortable with it. At 0500 I marked us back in service (available to take runs) while everybody else was snuggled in bed. Nagging doubt caused me to toss and turn, losing the little bit of precious sleep time left. Being my own boss on my days off, I decided to sleep for a few hours at home before I started cleaning an empty rental.

The next firehouse day, the first ring of the Centrex was for me. The lieutenant in the EMS office told me there would be an official, state investigation into the botched shocking to determine if it was machine or human error.

Hanging up the phone, if it was human error, I wondered what would happen to me. Will they give me remedial training? Yank my paramedic license? Blame the patient's death on me? (He died a day after he was transported to the hospital despite all our efforts.) I was as jumpy as a frog on a hot griddle. After a week the Centrex rang for me again. The state report was in. The motherboard in the defibrillator failed. Moisture had crept in and affected it. The lieutenant lectured me on how to care for the unit; always keep it in the cab, never in the compartment where it would be subjected to cold and heat. I cut him short, reminding him that my original equipment was being repaired because of printer problems. The unit that failed was a loaner. I couldn't be responsible for its treatment before I got it. At our station we keep the sensitive gear in the cab. He confessed that mine was not the first failure. He felt the answer to the problem was to get new gear. Ours was at its life expectancy and beginning to fail. Not wanting to take chances with people's lives, a few months later, the department ordered new defibrillators.

Our ex-Olympian weightlifter lived six months as predicted, dying as skinny as a wraith. Our crew attended the funeral. I will remember him laughing, calling me Princess as he walked away in his shorts and flip-flops with a lounger in his arms, going out behind the firehouse to sunbathe. He instilled in his children a love of life. His son, following in the footsteps of his father, was hired recently. He reminds me of his dad. They smile just alike, both great guys.

As I approach my 20th year of seeing so many horrors and losing tons of sleep, I feel I should be immune to events that cause nightmares. But it seems there is always something that will twist the knife, renewing the pain and sleepless nights.

We were recently dispatched to an injured person, being cautioned to take care, because the scene was not secure. We await the arrival of the police on those runs. Since the police beat us there, we immediately entered a small Mexican restaurant. The frightened, huddled people pointed us next door to the Mexican grocery store. At the back, by the service door, lay an old, heavyset woman, in a pool of blood. Her blood, spilled around her, was beginning to congeal. We turned her over and cut her shirt down the middle, revealing four to five, gaping, stab wounds. A neighboring fire department that we often go on runs with, and the ambulance crew arrived, which gave us a total of three paramedics on the scene.

Her skin was warm as I placed the electrodes from the monitor on her chest and distended belly. The monitor showed a flatline. I made a command decision not to work her. This was the first time I had to make that kind of decision. I knew in my head, it was the right choice. The other paramedics shook their heads in agreement, patting me on the back, reassuring me. But my heart didn't agree. An ever hopeful, optimistic voice said we should at least try a few chest compressions and start an IV to administer the heart starting drugs. As we packed up, I noticed the cut marks on her hands where she fought against her attacker. I shook my head in disgust at the senseless loss of life, probably for a few dollars in the cash register, drug money for some crackhead.

Outside, in our little circle, after we left the murder scene as undisturbed as possible, everyone reassured me that I did the right thing. Despite all the reassurance, that night, every time I shut my eyes to go to sleep, her face flashed in my mind. Needing sleep because I had a busy day scheduled, but afraid to close my eyes, I got up and turned on the TV, sitting for hours in the recliner, bleary-eyed, mindless, dozing off at some point.

Another time we were sent to a trailer park for a sick person. The ambulance arrived right behind us. We entered the trailer meeting some neighbors who pointed us to the bedroom. An older man, maybe in his 60s, was sitting on the edge of his bed, staring at a bowl of soup with crackers crushed in it. He said he appreciated his thoughtful neighbors for cooking it, but he just wasn't hungry. After taking his blood pressure, we informed him it was sky-high. We asked him which hospital he wanted to go to, but he said he wasn't going to any hospital. Apprising him he needed to go or he might die, he still stubbornly refused.

A protracted discussion ensued with us arguing why he needed to go and him countering with why he wasn't going. He said he didn't have the money. We said that didn't matter. He said it would make him sicker to get bills he could not pay. His daughter, who lives out of town, spoke to us long distance. She was the one who called the neighbors and 911. We told her we would be happy to take him to the hospital, but he refused. This being America, he has the right to say no to medical treatment. If we take him against his will, it would be kidnapping. She pleaded with us, but there was nothing we could do without his permission since he was lucid and aware of his actions. The neighbor witnessed his signature. Our forms state that a person could die or be disabled if they don't go to a doctor. It informs them they can call us back any time by dialing 911. It warns them they are assuming full responsibility for their own care. He signed our papers releasing us from liability.

Two work days later, I got a call from the EMS chief. He wanted the original paperwork from the run when the patient refused to go. The chief wanted to know why I let this guy sign a statement of release. I discussed his adamant refusal to go. He told me the patient died two days after he signed our papers. The daughter was mad as hell because we hadn't taken him in. After we left, she tried to reach her father for days. She convinced the police to break in. They found his bloated body.

My old paramedic instructor couldn't have given me a better lesson. I finally understood his vehemence in forcing people to go to the hospital against their will. It would have saved this man's life. Perhaps if I had been more forceful, he would have lived to see his daughter at Christmas dinner.

My instructor's words, "You can sue me tomorrow, because you will still be alive." rang in my ears.

Since I had carefully documented, in his own words, this patient's refusal to go to the hospital, I would not get in any trouble or face a lawsuit. Did I feel good about this? No! It stinks having shit like this happen! Next time, I think I will pull out my knife and threaten to bring the patient in alive.

My academy classmates started talking about getting together for our 20th anniversary of fire department service. How could it be that long? It doesn't seem like it.

Since we are spread over three shifts, it is impossible to get everybody together at once. We would need two parties for everyone to attend and three for everyone to see each other. In the end, the day slipped by with no parties just like the end of our probationary year. Our respective chiefs brought us our 20 year pins. Some classmates wear them on their jackets to education classes.

Oh, by the way, the grinding work of paramedic training is definitely worth it. Why did I wait so long?

The best 20 year gift I got was from classmate Mary who worked overtime at my station. She left a yellow sticky note with a handwritten message stuck to the big mirror in the women's locker room.

"We did it, girl, 20 years! Congratulations!!!"

Yes, we did. We are still doing it.

Afterword

This book, a long undertaking, is being published in my 22nd year of firefighting. My current station is a beautiful, quaint firehouse, Station 32, in the heart of Broad Ripple. I am on a squad on the A shift practicing my advanced life support skills as a paramedic. On fires, which are few but sometimes whopping, I run in with my axe or assist with hose lines. Squad work is varied, much like truck work, for which I have a fondness.

Come to Indianapolis. Visit Broad Ripple. It is a cultural mixture of restaurants, small shops, artist studios, coffee houses, book stores and good time bars. Check out my fire house. It is right next to the canal and the Monon Trail which is an old train track converted to a jogging trail for exercise enthusiasts in the area. Take a walk with me. Ever the gossip, I still have a few, good stories I can tell you.

Author's Note In the Interest of My Ex-husband

My ex asked me to put a note in the book for him. He swears he never cheated on me. Thinking of the evidence I documented in the book, it is difficult to believe him. If he is telling the truth, then I lived a tragedy of errors and many people suffered because of what I held to be true all these years, first and foremost, our son.

Glossary

Excerpted from my academy training manual
General Fire Service Terminology

Aerials- trucks which have an aerial ladder on them
Alarm- signal indicating need for fire department services
Apparatus- any squad, engine, or truck type vehicle
Arson- willful or malicious burning of property
Attack- any action to control fire
Backdraft- explosion or rapid burning of heated gases in a confined structure, usually caused by improper ventilation
Backstep- back end of an apparatus upon which firefighters stand and ride
Backstepper- firefighters who ride the back end of the apparatus
BLEVE- boiling liquid expanding vapor explosion
Booster line or Red line- small diameter hose handline used for small fires carried on reels on the apparatus
Box- type of alarm which calls for specific numbers and types of apparatus to be sent
Bunker clothes/turnout gear/protective clothing- firefighters' protective clothing
Butt- coupling of a fire hose or the bottom of a ladder
Catch a hydrant or plug/hooking up- to connect a hydrant to an engine
Charged line- a pressurized hose line ready to use
Chauffeur- a title used prior to Engineer, meaning the person who drives the apparatus
Confinement- operation to prevent spread of fire to uninvolved areas
Control- fire is under control when spread is halted

Drill tower- a place where realistic practice can be obtained
Engine- a pumper
Engine company- a group of firefighters equipped with a pumper
Engine house/quarters- any fire station
Engineer- the person who drives the apparatus
Evolution- a sequential operation of fire service training
Exposure- Property endangered by a fire
Extinguisher or can- a portable firefighting appliance filled with water or other extinguishing substance
Fill in- to move an apparatus from one station to another because the area equipment has been sent to a large fire
District- a geographical area dividing the city into four parts
Fire gases- gases produced as combustion occurs
Fireground- an area around a fire occupied by firefighting forces
Folding/attic/scuttle hole ladder- narrow ladder used for getting into tight spaces
Flashover- when contents of a room are heated to the point that flames flash over the entire area
Fly- top sections of a ladder raised by a rope
Fog- finely divided particles of water used for fire control
Foot a ladder- to hold the bottom of a ladder to the ground while it is being raised or lowered
Fully involved- when an entire area is completely involved with heat, smoke and flame
General order- a standing order distributed to all members of the department
High rise- a building that requires fire fighting tactics other than ground based operations
High value district- section of a city in which valuable property is located

Hose tower- part of a station designed so hose can be hung vertically to drain and dry
Hydraulics- study of use and movement of fluids at rest and in motion
Jake brake- motor retarder which is activated upon release of the gas pedal
Kelly day- a day off that was given to firefighters in the past but no longer given on this department
Knock down- to bring a fire under control
Ladder the building- to place ladders where needed, windows, roofs, all sides of the building
Layout- hose from an engine distributed at a fire
Maltese cross- symbol of a firefighter worn on uniform
Mutual aid- reciprocal assistance rendered to other fire departments in time of an emergency
Nozzle- device at the end of a hoseline to give shape, velocity and direction to the water or fire stream
Open up- to ventilate a building or use forcible entry
Pole- the sliding pole from upper stories of a fire station
Preconnected- when hose is connected to inlet or outlet before a fire to expedite hose operations
Pump panel- instrument panel on engines from which the engineer controls and monitors the pumps
Raise/throw- the way a ladder is stood up and placed in service
Rekindle- reignition of a fire by sparks, heat or embers, proper overhaul prevents this
Reserve apparatus- equipment used when the first-line apparatus is out of service for repairs
Run- a response to an emergency
Seat of fire- area where main body of fire is located
Shop- fire department maintenance and repair area
Shoulder carry- method of loading hose so that it can be advanced while carrying on the shoulder
Size-up- mental process of evaluating all the factors at a fire prior to commitment of personnel and equipment

Substitute/sub- term for firefighters who are new to the department and who go from station to station replacing regular firefighters who are off

Suppression- activities designed to overcome and extinguish a fire

Through the roof- visual indication that a fire has vented itself by burning a hole through the roof

Truck company- a group of firefighters equipped with a ladder truck

Ventilation- Permitting or forcing heated smoke and gases of a fire to escape to the atmosphere

Watch- period of time during which a firefighter is assigned to the communication center in the station

Working fire- a fire at which considerable fire fighting activity is being conducted

IFD Slang Terms

Also excerpted from my training academy manual

Box company- 3rd engine listed on a box alarm or any apparatus not shown on the still but on the box alarm
Can- extinguisher
Catch the run- respond on the apparatus
Catch the seat- assume the duties of the officer in the officer's absence
Catch the wheel- assume the duties of the engineer
Centrex- fire department telephone- official business
Charge the line- turn the water on to the hose
Crispy Critter- fire fatality
Croppie- person who has died
Didn't turn a wheel- no runs during 24 hour shift
Fill the bell- Christmas fund raising for needy children
In service- available for a run
Mask- self contained breathing apparatus SCBA
On station- apparatus is at the station
Out of service- not available for a run
Pick up- shut water off to the hose, drain and put it back on the apparatus, help other companies
Square wheels- an apparatus that isn't dispatched often
Tank- air or water tank
Trade time- a firefighter gains permission to have another firefighter work all or a portion of his shift then is obligated to repay her in the future. It is illegal to pay a firefighter to trade time.
2nd chauffeur- person designated to drive in engineers absence

About the Author

Born in 1954 and raised in Indianapolis, I, Kathy Gillette, left for a few years while serving in the U.S. Army in Germany. I missed home and came back to major in Physical Education and play basketball at Indiana University. After a knee injury during basketball practice, I switched majors and earned degrees in German and Business Management, taking five years to finish my college education. Unsatisfied with the business world and hearing my brother's adventures on the fire department, I decided to apply. After a lengthy process, with thousands of applicants, I was hired in March of 1985 along with 31 other people. Twenty-two years later, there are 21 of us from my class still fighting fire. I signed papers announcing my retirement effective March 9, 2010. On my days off, I manage a successful rental business. My son is grown, knocking heads as a bouncer. My husband is a police officer and we have his two teenage daughters at home.

Look for

"Happy Sad Firefighter Stories"
Another great book coming soon
From Kathy Gillette and
Alacheri Publishing, LLC

You may place an advance order for "Happy Sad Firefighter Stories" by sending a check or money order for $26.95 + $5.00 for shipping = $31.95 to
Alacheri Publishing, LLC
PO Box 26587
Indianapolis, IN 46226

CD version available

Order your copy of "Firefighterette Gillette" on CD by going to: www.firefighterettegillette.com or
www.alacheripublishing.com or

Exercise Videos are also available

Firefighterette Gillette Introduction to Exercise
Walk for Fun

Firefighterette Gillette Beginning Exercise
Dance for Fun

Firefighterette Gillette Water Exercise
Swim for Fun

Firefighterette Gillette Intermediate Exercise
Expert Fun

Firefighterette Gillette Advanced Exercise
Hard but Fun

About the cover picture

On Saturday, March 4, 1989, my company and several others were dispatched to the report of a house fire. As we were searching for the seat of the fire in a dark, smoky, double residence, a back draft occurred with an eruption of flames so sudden that it left nothing but orange all around my buddies and burned or injured all of them. The explosion blasted them backward down the steps where they fell in a heap. Being the new guy, I had been sent to get some equipment on the truck and was spared from the blast. I assisted them outside and into ambulances. After they left, I was distraught and worried because one of them was hurt so badly, I thought he might never return to work. Two officers were trying to reassure me that nobody was going to die and probably everyone would be back to work in a few weeks or so. The picture was snapped as I was listening, wiping the sweat, dirt, and blood off my face with a shaky hand.

Notes

Notes

Notes

Notes

Notes

Notes

Notes

Notes

Notes

Notes